AF588196

Current Topics in Behavioral Neurosciences

Volume 24

About this Series

Current Topics in Behavioral Neurosciences provides critical and comprehensive discussions of the most significant areas of behavioral neuroscience research, written by leading international authorities. Each volume offers an informative and contemporary account of its subject, making it an unrivalled reference source. Titles in this series are available in both print and electronic formats.

With the development of new methodologies for brain imaging, genetic and genomic analyses, molecular engineering of mutant animals, novel routes for drug delivery, and sophisticated cross-species behavioral assessments, it is now possible to study behavior relevant to psychiatric and neurological diseases and disorders on the physiological level. The *Behavioral Neurosciences* series focuses on "translational medicine" and cutting-edge technologies. Preclinical and clinical trials for the development of new diagnostics and therapeutics as well as prevention efforts are covered whenever possible.

More information about this series at http://www.springer.com/series/7854

David J.K. Balfour · Marcus R. Munafò
Editors

The Neuropharmacology of Nicotine Dependence

Editors
David J.K. Balfour
Ninewells Hospital
University of Dundee Medical School
Dundee
UK

Marcus R. Munafò
School of Experimental Psychology
University of Bristol
Bristol
UK

ISSN 1866-3370 ISSN 1866-3389 (electronic)
Current Topics in Behavioral Neurosciences
ISBN 978-3-319-13481-9 ISBN 978-3-319-13482-6 (eBook)
DOI 10.1007/978-3-319-13482-6

Library of Congress Control Number: 2014956492

Springer Cham Heidelberg New York Dordrecht London

Printed on acid-free paper

Springer International Publishing AG Switzerland is part of Springer Science+Business Media (www.springer.com)

Preface

When one of us (DJKB) first started studying the psychopharmacology of nicotine some 40 years ago the numbers of researchers interested in the topic was small and could probably be accommodated around a large dinner table. Our understanding of the potential hazards of smoking was at a fairly early stage as was our understanding of the neural mechanisms that mediated the behavioral responses to nicotine. At that time smoking was considered to be a habit, not an addiction, and was still widely accepted. Readers who are not old enough to remember those times may be familiar with the television series, *Mad Men*. That series gives you an impression of how acceptable smoking was. Even into the 1980s, the fact that neurones within the brain expressed nicotinic receptors was still debated among some researchers. We have come a long way since that time, and now it is not unusual to have 1,000 delegates or more at conferences on nicotine and tobacco, and sessions dedicated to nicotine are not uncommon at many neuroscience conferences. Moreover, public health policy is now driven by a sound evidence base relating both to the toxicity of primary and second-hand (also known as environmental) tobacco smoke and the plethora of neuroscience studies that have established nicotine as one of the most widely studied recreational drugs. The primary purpose of the chapters in this book and its companion volume is to explore the extent to which the wide range of approaches adopted to investigate the behavioral responses to nicotine and the molecular and neural mechanisms that mediate these effects have opened our eyes to the properties of this unique and fascinating drug.

It goes without saying that one of the principal factors that drives the study of nicotine psychopharmacology is its established role in the addiction to tobacco. It is appropriate, therefore, that this second volume is dedicated specifically on this issue. The chapters in this volume not only describe the ways in which research at a basic level, largely using animal models, have revealed the complex mechanisms that seem to underpin the role of nicotine in tobacco smoking, but also the ways in which the results of these studies translate to our understanding of the dependence on tobacco experienced by most habitual smokers. A number of the chapters show how modern imaging technologies have allowed us to relate directly findings in animal models to the effects of nicotine and tobacco smoke in the human brain.

We have sought to take a logical approach to the issue by first addressing the neurobiological and psychological mechanisms that contribute to the rewarding, perhaps better called the reinforcing, properties of nicotine. We then turn to the mechanisms that underpin the effects of nicotine withdrawal and relapse, chapters that will have a particular resonance with smokers. The final chapter returns to the issue of the role of underlying psychiatric illnesses in tobacco dependence. It focuses on the ways in which animal studies have contributed to our understanding of the reasons that this group seems to be especially vulnerable to tobacco dependence and resistant to treatment.

We hope that the volumes *The Neurobiology and Genetics of Nicotine and Tobacco* and *The Neuropharmacology of Nicotine Dependence* will provide readers with a contemporary overview of the current research on nicotine psychopharmacology and its role in tobacco dependence from leaders in this field of research and that they will prove valuable to those who are developing their own research programs in this important topic.

Dundee, Scotland David J.K. Balfour
Bristol Marcus R. Munafò

Contents

Imaging Tobacco Smoking with PET and SPECT

Kelly P. Cosgrove, Irina Esterlis, Christine Sandiego, Ryan Petrulli and Evan D. Morris

Abstract Receptor imaging, including positron emission computed tomography (PET) and single photon emission computed tomography (SPECT), provides a way to measure chemicals of interest, such as receptors, and neurotransmitter fluctuations, in the living human brain. Imaging the neurochemical mechanisms involved in the maintenance and recovery from tobacco smoking has provided insights into critical smoking related brain adaptations. Nicotine, the primary addictive chemical in tobacco smoke, enters the brain, activates beta2-nicotinic acetylcholine receptors (β_2*-nAChRs) and, like most drugs of abuse, elicits dopamine (DA) release in the ventral striatum. Both β_2*-nAChRs and DA signaling are critical neurosubstrates underlying tobacco smoking behaviors and dependence and have been studied extensively with PET and SPECT brain imaging. We review the imaging literature on these topics and describe how brain imaging has helped inform the treatment of tobacco smoking.

Keywords Brain imaging · Smoking · Nicotine · Nicotinic acetylcholine receptors · Dopamine · PET · SPECT

K.P. Cosgrove (✉) · I. Esterlis · C. Sandiego · E.D. Morris
Department of Psychiatry, Yale University School of Medicine,
2 Church Street South, Suite 511, New Haven, CT 06519, USA
e-mail: kelly.cosgrove@yale.edu

K.P. Cosgrove · I. Esterlis · E.D. Morris
Department of Diagnostic Radiology, Yale PET Center,
Yale University School of Medicine, 2 Church Street South,
Suite 511, New Haven, CT 06519, USA

R. Petrulli · E.D. Morris
Department of Biomedical Engineering, Yale PET Center,
Yale University School of Medicine, 2 Church Street South,
Suite 511, New Haven, CT 06519, USA

D.J.K. Balfour and M.R. Munafò (eds.), *The Neuropharmacology of Nicotine Dependence*, Current Topics in Behavioral Neurosciences 24,
DOI 10.1007/978-3-319-13482-6_1

Contents

1 Introduction

Positron emission computed tomography (PET) and single photon emission computed tomography (SPECT) are unique among imaging techniques in the ability to measure specific molecules in the brain. They have revolutionized our ability to measure chemicals in the brains of living people. Brain chemicals, including receptors present in low concentrations (nM–pM range), are measured using "trace" doses of highly specific radioactive drugs (called radiotracers) and imaged with a PET or SPECT camera. There are differences in the physics and chemistry used in PET versus SPECT, but the outcome—a measure of receptor availability—is the same (Fig. 1). The primary addictive chemical in tobacco smoke, nicotine, activates β_2*-containing nicotinic acetylcholine receptors (β_2*-nAChRs). Although there are other combinations of subunits that assemble to form nAChRs (see chapter entitled Structure of Neuronal Nictinic Receptors; volume 23), the β_2*-nAChRs are pivotal (e.g., they are critical for the reinforcing effects of nicotine), and so this site has been a primary point of interest for radiotracer imaging in the smoking field. In addition, when nicotine activates β_2*-nAChRs located on mesolimbic dopamine (DA) neurons in the ventral tegmental area, this results in neuronal firing and dopamine release in the nucleus accumbens (Imperato et al. 1986). There are several SPECT and PET radiotracers that label β_2*-nAChRs and are used to measure occupancy and changes in receptor availability, and several radiotracers that label dopamine D2/3 receptors and allow for measurement of fluctuations in synaptic dopamine. Both β_2*-nAChRs and DA signaling are critical neurosubstrates underlying tobacco smoking behaviors and dependence. In this chapter, we will review the imaging literature that has provided insights into the molecular mechanisms of tobacco smoking with a focus on studies examining β_2*-nAChR availability and DA neurotransmission.

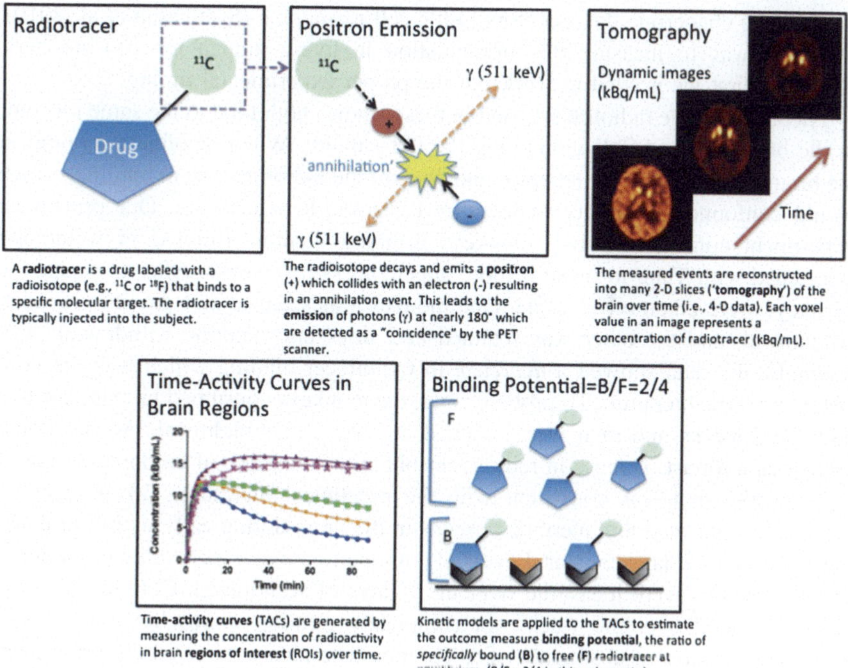

Fig. 1 Description of the use of PET brain imaging to determine the outcome measure of receptor availability

2 Imaging β2*-Nicotinic Acetylcholine Receptors

2.1 Preclinical Studies

The nAChRs that contain the β_2*-subunit are critical for mediating the effects of nicotine in the brain including the reinforcing effects (Picciotto et al. 1998), dopamine release (Epping-Jordan et al. 1999; Koranda et al. 2013), sensitivity to nicotine (Cosgrove et al. 2010; Marubio et al. 1999; Tritto et al. 2004), and the incentive aspects of motivation (Brunzell et al. 2010) (see chapter entitled The Role of Mesoaccumbens Dopamine in Nicotine Dependence; this volume). In addition, there is a wealth of literature showing that nicotine and tobacco smoking robustly upregulate (i.e., increase numbers of) β_2*-nAChRs throughout the brain (Abreu-Villaca et al. 2003; Benwell et al. 1988; Breese et al. 1997; Kassiou et al. 2001; Marks et al. 1992). Preclinical studies administering nicotine at various doses and routes of administration to rats, mice, and monkeys—as well as postmortem human studies—have all indicated that nicotine and tobacco smoke result in significantly more β_2*-nAChRs throughout the brain compared to saline (animals) or to not smoking (humans). We now know that nicotine itself is responsible for this upregulation. Nicotine acts in the cell to help the receptor subunits assemble and then acts to chaperone the receptors to the cell membrane (Srinivasan et al. 2010). Our goal was to measure this upregulation in living human tobacco smokers. However, first we needed to work out the proper experimental timing.

Nicotine and the radiotracers used in these studies both bind to the same receptor in the brain—the nAChR containing the β_2*-subunit. When nicotine is present in the brain, it may block the receptor and prevent the radiotracer from binding, which would confound our ability to measure β_2*-nAChR availability. Our preclinical experiment consisted of two monkeys drinking nicotine (diluted in water and sweetened with Tang to make it more appetizing) for 6 weeks. After 6 weeks, the monkeys were taken off nicotine. One monkey was scanned at 1 day into nicotine withdrawal, and the other was scanned at 2 days into nicotine withdrawal. Surprisingly, the data showed a *decrease* in radiotracer binding which was not consistent with the literature. To probe further, the monkeys consumed nicotine for two additional weeks and then were scanned at 7 days of withdrawal. At that point, there was a robust increase in radiotracer binding suggestive of an upregulation of β_2*-nAChRs that was consistent with the preclinical literature. Taken together, these data suggested that nicotine remains in the brain during early withdrawal and may take up to 7 days to clear. Levels of cotinine (the major metabolite of nicotine) in the monkeys were measured over the 7 days of withdrawal. Cotinine progressively declined over the week, not completely clearing or reaching nonsmoker levels until 7 days of abstinence. The cotinine data nicely mirrored the brain data. Once cotinine had cleared, it was possible to measure nicotine-induced upregulation of β_2*-nAChRs in the brain. In the human studies discussed below, low cotinine levels are typically used as an indicator of abstinence and that nicotine has cleared so that smokers can be imaged with β_2*-nAChR radiotracers.

2.2 *Imaging the Upregulation of β_2*-nAChRs in Tobacco Smokers*

Based on the preclinical monkey study, our group imaged β_2*-nAChRs in human tobacco smokers at 7–9 days of smoking abstinence (early phase withdrawal). In this and similar studies, the subjects were required to quit smoking and not use any medications or nicotine replacement strategies such as the nicotine patch, because all forms of nicotine would bind the β_2*-nAChR and block the radiotracer from binding. In order to help the subjects quit smoking, we used contingency management techniques (Staley et al. 2006). In the first paper, we demonstrated that tobacco smokers at 7–9 days of abstinence have significantly higher β_2*-nAChR availability in the cortex, striatum, and cerebellum compared to a group of age- and sex-matched nonsmokers (Fig. 2). This work in our laboratory (Staley et al. 2006) and others (Mamede et al. 2007; Mukhin et al. 2008) confirmed that it is possible to measure the upregulation phenomenon in human smokers in vivo.

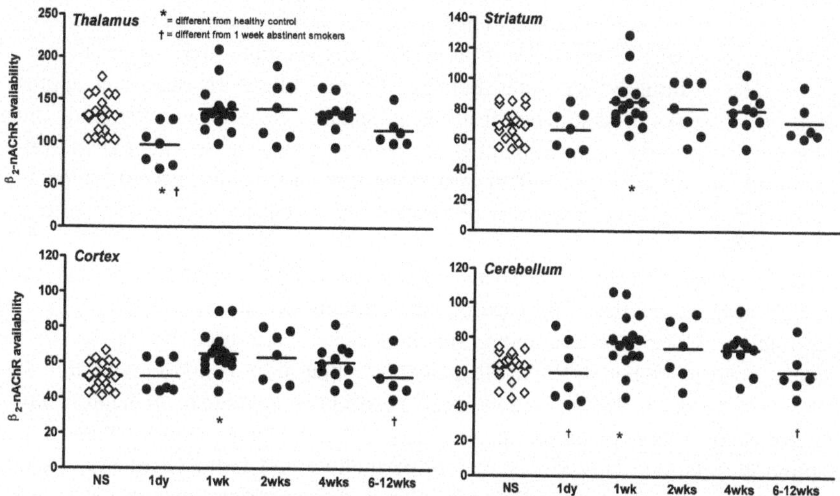

Fig. 2 β_2*-nAChR availability (V_T/f_P) is shown in individual nonsmokers (*open diamonds*) and tobacco smokers (*filled circles*) at 1 day, 1, 2, 4, and 6–12 weeks of abstinence in the thalamus, striatum (average of caudate and putamen), cortex (average of cortical regions including parietal, frontal, anterior cingulate, temporoinsular, and occipital cortex), and cerebellum. The line in each scatter plot represents the mean value of those subjects. *Asterisk* indicates significant difference from control nonsmokers after Bonferroni's correction using two-sample *t*-tests. *Dagger* indicates significant difference from 1 week abstinent smokers after Bonferroni's correction using planned post-hoc between-group comparisons subsequent to the analysis of repeated measures mixed-effects regression models including the overall effect of abstinent smoker group

2.3 *Imaging the Normalization of β_2*-nAChRs in Tobacco Smokers*

There is also evidence from preclinical (Collins et al. 1990; Pietila et al. 1998) and postmortem human (Breese et al. 1997) studies that the β_2*-nAChRs do not stay upregulated and eventually return to control levels. The postmortem study (Breese et al. 1997) indicated that smokers who had quit smoking at least two months prior to their death had β_2*-nAChRs levels similar to controls. However, smokers in the study had quit anywhere from 2 months to 30 years prior to their death, so the study did not shed light on the acute time course of receptor changes (e.g., during acute withdrawal in the first few months of abstinence, when relapse rates are high). In our next study, we imaged β_2*-nAChR changes over the first few months of abstinence in tobacco smokers (Cosgrove et al. 2009). As shown in Fig. 2, at one day of abstinence, nicotine is still present in the brain blocking the radioligand from binding to the receptor and there is no difference in β_2*-nAChR availability compared to the group of nonsmokers. At one week of abstinence, there is higher β_2*-nAChR availability in smokers compared to nonsmokers consistent with the previous study (Staley et al. 2006). Then even at 2 and 4 weeks of abstinence, receptor availability remains high and does not return to nonsmoker control levels until 6–12 weeks of abstinence. This study (Cosgrove et al. 2009) and others (Brody et al. 2013b; Mamede et al. 2007) demonstrate that upregulation of β_2*-nAChRs is initially persistent, but that β_2*-nAChRs normalize over approximately 6–12 weeks of abstinence from cigarettes and all other nicotine-containing products. These brain changes parallel the clinical course of smoking cessation in which craving, relapse, and withdrawal symptoms slowly dissipate over the first few months of abstinence even though relapse may occur months or years after the last cigarette.

Relationships between β_2*-nAChR availability and clinical correlates have been reported in these studies. The type of cigarette smoked modulates the degree of upregulation. Cigarettes containing menthol, which are used by up to 1/3 of smokers, lead to higher β_2*-nAChR availability than non-menthol-containing cigarettes (Brody et al. 2013a). Additionally, a primary advantage of neuroreceptor imaging studies (vs. postmortem studies) is that we can record behavior and examine correlations between behaviors of interest and brain chemistry. At one week of abstinence, β_2*-nAChR availability in the sensorimotor cortex was negatively correlated with the urge to smoke to relieve withdrawal symptoms (Staley et al. 2006). At four weeks of abstinence subjects with higher β_2*-nAChR availability in the cerebellum reported both a greater desire to smoke and a greater urge to smoke to relieve withdrawal (Cosgrove et al. 2009). This suggests that magnitude of upregulation may play a role in craving over the course of abstinence and that managing the time course of the normalization may help individuals who are more likely to relapse in response to high levels of craving. For example, it is possible that nicotine replacement strategies may be effective because they continue to activate β_2*-nAChRs. This leads to continued upregulation and the individual can "wean" the receptors off of nicotine as the dose of nicotine is decreased over time.

2.4 Sex Differences in β_2*-nAChR Availability

There is a large literature demonstrating sex differences in tobacco smoking behaviors. In general, men tend to smoke for the nicotine reinforcement, or nicotine per se in the cigarette, whereas women tend to smoke more for the sensory cues associated with smoking, as well as affect and stress regulation (Perkins 2009; Perkins et al. 1999; Perkins and Scott 2008). There are also two preclinical studies showing that male rats and mice exposed to nicotine exhibited greater nAChR upregulation than female rats and mice exposed to nicotine (Koylu et al. 1997; Mochizuki et al. 1998). We wanted to determine if there were sex differences in β_2*-nAChR availability between men and women smokers compared to nonsmokers. Consistent with the preclinical literature, male smokers had significantly higher β_2*-nAChR availability compared to male nonsmokers (9–17 %), but women smokers had similar β_2*-nAChR availability compared to women nonsmokers (1–3 %) (Cosgrove et al. 2012). This was a striking finding given in all the studies demonstrating that nicotine and tobacco smoking upregulate β_2*-nAChRs throughout the brain. Considering known behavioral sex differences in tobacco smoking, these findings make sense and provide a biological mechanism that may underlie some of the behaviors. Specifically, men smoke for the nicotine in cigarettes, they are more responsive to nicotine replacement therapy as a cessation strategy, and men's brains are responsive to nicotine, exhibiting upregulation of β_2*-nAChRs. Women smoke for affect regulation and for reasons other than the nicotine, they do not respond as well to nicotine replacement strategies, and their brains do not respond to nicotine by increasing β_2*-nAChRs. The bottom line is that novel treatment strategies targeting other receptor systems need to be evaluated to more effectively help women quit smoking. All the current strategies act at the β_2*-nAChR, and, of course, all nicotine replacement strategies act at that site. For example, varenicline (Chantix) is a partial agonist at the β_2*-nAChR, and even bupropion (Zyban) is a nicotinic antagonist.

2.5 Nicotine Occupancy of β_2*-nAChRs

In addition to receptor changes, imaging studies have informed our knowledge about what happens in the brain after someone smokes a cigarette. For example, after one puff of a cigarette, approximately 50 % of all β_2*-nAChRs in the brain are occupied by nicotine. After smoking one or two cigarettes, the receptors are saturated, so up to 100 % of β_2*-nAchRs are occupied by nicotine (Brody et al. 2006a; Esterlis et al. 2013). We know that nicotine doesn't *clear* the brain immediately, and in fact dependent smokers have a slower process of nicotine accumulation going *into* the brain from a cigarette than do nondependent smokers (Rose et al. 2010). Indeed, in one study, habitual smokers did not show evidence of puff-associated spikes in nicotine, but rather a gradual accumulation of nicotine during smoking (Rose et al. 2010). Both of these ideas—rapid accumulation and puff-associated spikes of nicotine—had been proposed to explain the maintenance of tobacco dependence. So with a slow kinetic profile and with most receptors in a smoker occupied by nicotine throughout the day, why do people keep smoking? This brings up some important points about tobacco smoking. People smoke for many different reasons, and nicotine reinforcement is only one component. The reinforcement or pleasure derived from nicotine, like many other drugs of abuse, is necessary in driving the initial phases of drug-seeking behavior. However, as the addiction progresses, many people may continue to smoke in order to avoid withdrawal symptoms and due to the many conditioned cues that have become ingrained, which are a part of the compulsive, repetitive nature of tobacco smoking. Additionally, there are over 4,000 chemical compounds that are produced when a cigarette burns; all of these compounds are in tobacco smoke and are inhaled. Thus, while nicotine is the primary addictive component of tobacco smoke, there are additional compounds such as MAO-A and MAO-B inhibitors that likely play a role.

Other imaging studies have demonstrated that even smoking a denicotinized cigarette, which supposedly has very low nicotine content, leads to occupancy of approximately 20 % of β_2*-nAChRs in the brain (Brody et al. 2009). This is similar to the level of occupancy produced by secondhand smoke. Brody and colleagues at UCLA performed an elegant study examining the effect of secondhand smoke on β_2*-nAChRs by having subjects sit in a car (the window was down a few inches) with a person smoking, and they reported up to 20 % of β_2*-nAChRs were occupied by nicotine in the individual who was just sitting in the car, not smoking (Brody et al. 2012). Interestingly, several states have recently passed laws prohibiting smoking in the car with children under the age of 18.

In terms of treatment, Esterlis and colleagues conducted a proof-of-concept study (Esterlis et al. 2013) to determine whether a nicotine vaccine, 3′-AmNic-rEPA, would reduce the amount of nicotine entering the brain and binding to β_2*-nAChRs. Smokers were imaged before and after treatment with the vaccine, and occupancy of intravenously delivered nicotine was measured. Immunization led to a significant ~13 % reduction in nicotine occupancy confirming that vaccines may help smokers quit smoking by reducing the amount of nicotine available to occupy β_2*-nAChRs. While this particular vaccine is not available for treatment, this study highlights an innovative paradigm to test future medications.

3 Imaging Dopamine Release in Response to Nicotine and Tobacco Smoking

3.1 Preclinical Microdialysis Studies

Before PET made possible indirect measurement of dopamine release in vivo, the only way to measure dopamine levels in a living brain was via microdialysis. Microdialysis is used primarily in rodents. During a surgery, a probe is placed through the skull into the region of interest (e.g., the nucleus accumbens). After recovery, DA levels can be sampled in the awake, behaving animal typically in response to a drug or a stimulus. Di Chiara and Imperato performed the seminal study showing that drugs abused by humans release DA in the nucleus accumbens of the rat brain (Di Chiara and Imperato 1988). Amphetamine (1.0 mg/kg, SC) raised DA levels over 1,000 % from baseline, whereas ethanol (1.0 g/kg, IP) and nicotine (0.6 mg/kg, SC) raised DA levels to 200 % over baseline levels. This illustrates how powerful DA release can be when it is directly stimulated with amphetamine, which is both a direct DA releaser and DA reuptake inhibitor. Put into context, a similar dose of amphetamine given to a monkey or a human in a PET experiment, which is an indirect measure of change in DA, would result in a 15–30 % change in binding potential (BP) as measured with [^{11}C]raclopride or another D2/3 ligand.

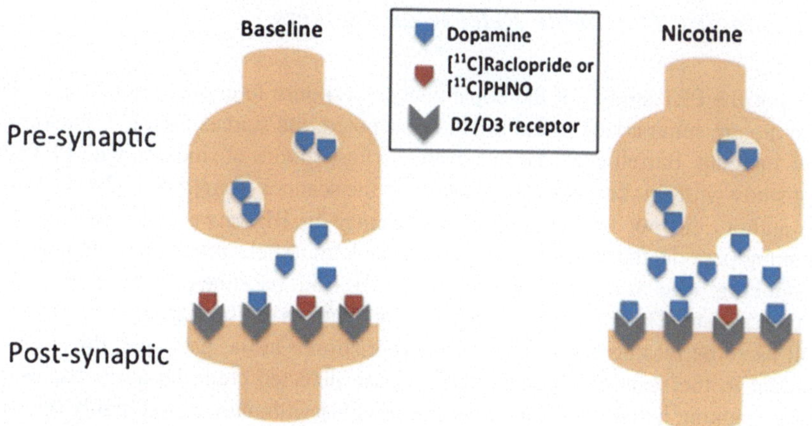

Fig. 3 Binding competition at dopamine D2/D3 receptors between endogenous dopamine and PET radiotracers, [^{11}C]raclopride or [^{11}C]PHNO, at baseline and after nicotine or tobacco smoking. Nicotine increases synaptic dopamine levels, and the dopamine binds to the D2/3 receptors which reduce the number of available binding sites for the radiotracers which also bind the D2/D3 receptors. The change in number of available receptors from the baseline scan to the scan after the drug challenge is an indirect measure of increased dopamine levels

3.2 *Imaging Dopamine Release with PET*

Striatal dopamine release has been reliably measured using stimulants such as amphetamine or tobacco smoking with radiotracers such as [^{11}C]raclopride PET or [^{123}I]IBZM SPECT (see Laruelle 2000 for review). More recently, [^{11}C]PHNO has been used, since as an agonist, it labels the high-affinity functional dopamine D2/3 receptors (which theoretically makes it more sensitive to changes in DA compared to [^{11}C]raclopride, an antagonist). Further, because it has a higher affinity for D3 versus D2 DA receptors, this allows for regional interpretation of D3 and D2 receptors (Girgis et al. 2011). Specifically, in humans, binding in dorsal striatum is primarily D2, binding in substantia nigra is primarily D3, and binding in globus pallidus is mixed with approximately 65 % D3 versus D2 dopamine receptors (Tziortzi et al. 2011). For all of the dopaminergic radiotracers, drugs such as amphetamine robustly increase synaptic DA. The increased DA competes with the radiotracer to bind at the dopamine receptor; thus, an increase in DA results in a decrease in radiotracer binding compared to baseline (Fig. 3). This allows calculation of the “occupancy” of the receptors by DA or a change in BP and is an indirect measure of DA release based on the “occupancy model” (Laruelle 2000). Although nicotine is a less robust DA releaser compared to amphetamine, there are many studies that have examined nicotine and tobacco smoking-induced DA release in human subjects.

3.3 Imaging Dopamine Release in Smokers

Most of the PET studies of the dopaminergic response to cigarette smoking and/or nicotine are summarized in Table 1. Unless noted, the studies used [^{11}C]raclopride PET imaging. Barrett et al. did one of the earliest studies of smoking with PET; the first study with smokers actually smoking in the scanner (Barrett et al. 2004). While the authors did not find any significant change in BP between the baseline and smoking periods, the subjects who experienced "mood elevating effects in response to smoking" ($n = 5$) had a 21 % decrease in BP from baseline (i.e., increase in DA) in the caudate. There were some notable innovations in the Barrett et al. study as well as some reasons for caution in interpretation. On the plus side, the smokers smoked in the scanner—thus any DA release detected could be attributed to the entire smoking behavior—something not possible with animals but highly relevant for medications development. The smokers were asked to smoke their own brand—another way of assuring that behavior in the scanner approximated subjects' smoking behavior. However, the protocol of smoking up to six cigarettes in an hour may have been aversive to some of the subjects and possibly contributed to the variability between subjects.

Scott et al. (2007) took a different approach to the experimental design of their study. They asked a slightly different question than in the other studies: namely, what role does the nicotine per se in cigarettes play in DA release? Smokers smoked a denicotinized cigarette, followed by a regular nicotine-containing cigarette while in the scanner. In the 6 smokers, there was no statistically significant decrease in BP from denicotinized to nicotine-containing cigarettes in any subregion of the striatum. They did find a relationship between greater decrease in BP in the ventral striatum from denicotinized to nicotine-containing cigarettes (i.e., greater dopamine release) and degree of dependence on nicotine (as measured by the Fagerström Test for Nicotine Dependence; FTND) (Scott et al. 2007).

Two other studies focused on measuring the nicotine (as opposed to cigarette) effect on DA release in humans. Montgomery and colleagues administered nicotine to subjects via nasal spray (Montgomery et al. 2007), and Takahashi and colleagues had subjects chew nicotine gum (Takahashi et al. 2008). Neither group found any significant change in BP due to nicotine in any individual striatal region. Takahashi and colleagues showed a significant decline in BP with nicotine administration in the striatum overall indicating a dose-dependent effect. Following up on Scott et al. (2007) and Barrett et al. (2004) both newer studies looked for correlations between the change in BP and behavior. Montgomery and colleagues found a correlation between "happiness" and decrease in BP in associative striatum indicating greater happiness with greater dopamine release. In agreement with Scott and colleagues, Takahashi and colleagues found voxels in the ventral putamen that showed a significant correlation between decrease in BP and FTND score.

Finally, several studies have measured DA release with the subjects smoking outside the scanner. Brody and colleagues imaged smokers before and after a "smoke break" compared to smokers who took a break but did not smoke.

Table 1 Summary of smoking studies

PET Study	Subjects	Stimulus	Control	Findings
Brody (2004)	20 Smokers, 1 scan each	50 min Post-injection, 10 subjects smoked 1 *cigarette* at 10-min break outside of scanner; scanned again for an additional 30 min	10 Subjects did not smoke at 10-min break during scan	26–37 % Reduction in BP in left ventral caudate, nucleus accumbens, and left ventral putamen in subjects that smoked
Brody (2006)	45 Smokers, 1 scan each	35 Subjects smoked a *cigarette* during scan using same design as above study	10 Subjects did not smoke at 10-min break during scan	8 % Reduction in BP in ventral caudate and nucleus accumbens in subjects that smoked
Barrett (2004)	10 Smokers, 2 scans each	1–6 *Cigarette(s)* 15 min in scanner prior to radiotracer injection	No cigarette prior to radiotracer injection	No change in BP between conditions in the striatum; reduction in BP correlated with hedonia
Scott (2007)	6 Smokers, 6 matched healthy controls, 1 scan each	2 *Denicotinized cigarettes* at 2 and 12 min post-injection and 2 *cigarettes* at 40 and 50 min post-injection	Baseline receptor binding in healthy controls	Significant activation in left ventral basal ganglia corresponding to a 10 % reduction in BP between conditions
Montgomery (2007)	10 Smokers, 1 scan each	Displacement with *nicotine nasal spray* administered 50 min post-injection during infusion study	38–50 min (rest) compared with 58–100 min (post-nicotine)	No change in BP in striatum; negative correlation between BP and subjective measures of amused and happiness in the associative and sensorimotor striatum
Takahashi (2008)	6 Male smokers, 6 matched healthy controls, 2 scans each	*Nicotine gum* given 1 h prior to injection, subjects continued to chew throughout scan	Taste-matched placebo gum given 1 h prior to tracer injection	7 % Decrease in striatal BP in smokers with nicotine gum; no change observed in nonsmokers
LeFoll (2014)[a]	10 Smokers, 2 scans each	*Cigarette* before scan	No cigarette before scan	12 and 15 % Reduction in D2 and D3 BP, respectively, with cigarette

[a] LeFoll (2014) study scans were performed with [^{11} C]PHNO. All other studies measuring nicotine- or smoking-induced dopamine release were conducted with [^{11} C]raclopride

The major finding was a large (up to 26 %), but variable, amount of DA release in the ventral striatum, which was statistically greater in the smokers who smoked than in those who did not. In a follow-up study, the same group used the same protocol two years later, and while they replicated the direction of the findings, the effect size was quite a bit smaller (Brody et al. 2006b) with changes in BP of approximately 8 %. In both studies, they found that the greater the reduction of craving for a cigarette from pre- to post-break, the greater the increase in ventral striatal DA release. The 2006 paper introduces genetic variation and its possible role in smoking. Subjects were genotyped for mutations in the genes that encode for dopamine receptors, dopamine transporter, or catechol-O-methyltransferase (an enzyme that breaks down catecholamines) proteins. Their results suggested that differences in amount of DA released (change in tracer binding) could be related to mutation status, and this in turn helps to explain some of the inherent variability in the BP numbers in both studies. An additional study by the same group measured smoking-induced DA release before and after counseling, bupropion or placebo treatment. While smoking-induced DA release was reduced in all groups, there was no difference between treatments and the reduction was attributed to the smaller puff volume recorded after treatment (Brody et al. 2010).

In the most recent study, Le Foll and colleagues imaged smokers with [^{11}C]PHNO before and after a cigarette and BP was reduced after smoking by 12 and 15 % in D2-rich and D3-rich regions, respectively (Le Foll et al. 2013). This is consistent with our preclinical study demonstrating that [^{11}C]PHNO may be more sensitive than [^{11}C]raclopride to measure small changes in DA release elicited by nicotine.

One concern with the existing studies is the timing of the dopamine response. Specifically, the response to smoking a cigarette is a transient increase in DA; however, the image analysis techniques in the studies use an average of all the data collected over 30 min to up to 2 h, which may significantly dilute measurement of a transient dopamine response. The known limitations of common methods of analysis in the face of transient DA release are discussed in the recent paper by Sullivan and colleagues (Sullivan et al. 2013). Thus, analysis techniques with improved temporal resolution may be better suited to more transient DA release. We have recently developed a visualization technique to display 4-dimensional results in a "dopamine movie", where DA changes can be viewed in bins of 3 min, which allows for transient responses to be captured. Dopamine movies can be viewed at http://www.jove.com/video/50358 (Morris et al. 2013). The sensitivity of dopamine movies to spatially limited and temporally brief dopamine changes is fully characterized in a recent paper (Kim et al. 2014). Taken together these studies confirm that smoking elicits ventral striatal dopamine release and is associated with a reduction of craving. Development of this new model with greater temporal resolution of dopamine fluctuations will provide a novel paradigm to determine what aspects of the dopaminergic signature are different between groups or are affected by medications.

4 Imaging the Comorbidity of Tobacco Smoking with Schizophrenia

While approximately 20 % of the adult American population smokes cigarettes, the prevalence of tobacco smoking is up to 4 times higher in individuals with schizophrenia (see chapter entitled Psychiatric Disorders as Vulnerability Factors for Nicotine Addiction: What Have We Learned from Animal Models?; this volume). The exploration of the effects of smoking on brain chemistry in psychiatric populations is less advanced than in otherwise healthy individuals, but some early preclinical studies, and more recent clinical studies, provide insight into potential mechanisms underlying the high comorbidity.

An initial postmortem study suggested that smoking may not lead to the same extent of upregulated β_2*-nAChRs in individuals with schizophrenia compared to healthy controls. Specifically, nonsmokers with and without schizophrenia had similar numbers of β_2*-nAChRs; however, smokers with schizophrenia had lower numbers of β_2*-nAChRs than healthy smokers (Breese et al. 2000). We performed in vivo imaging studies to attempt to replicate their finding and to determine if there were relationships between β_2*-nAChRs and clinical correlates of schizophrenia. In our first study, we demonstrated that smokers with schizophrenia have lower β_2*-nAChR availability compared to healthy smokers (Fig. 4) (D'Souza et al. 2012). However, in that study, we did not include nonsmoker comparison groups and thus we were not able to determine whether there is a difference in degree of upregulation of β_2*-nAChRs in smokers with and without schizophrenia. Recently, we

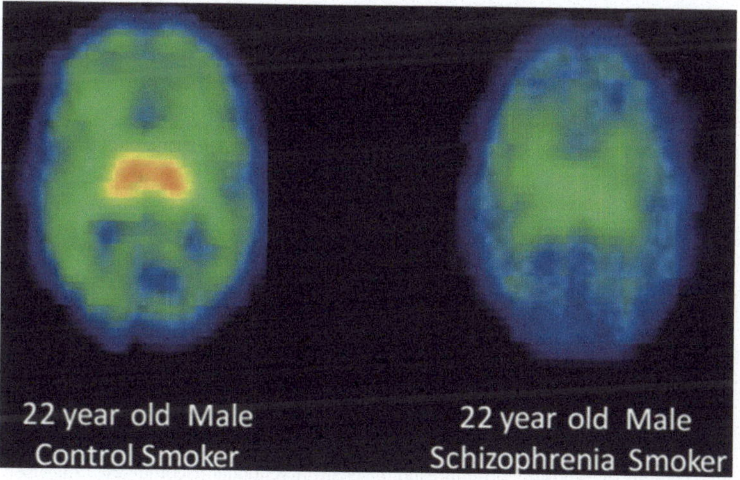

Fig. 4 Parametric images are shown illustrating β_2*-nAChR availability in a control male smoker (*left*) and a male smoker with schizophrenia (*right*) as measured with [^{123}I]5-IA-85380 SPECT. Higher receptor availability is observed in the control smoker (depicted by higher concentration of *red* and *yellow*) as compared to the smoker with schizophrenia

followed up with a larger study and examined this system in smokers and nonsmokers with and without schizophrenia. We confirmed the postmortem study and demonstrated that smokers with schizophrenia do have higher β_2*-nAChR availability than nonsmokers with schizophrenia; however, the difference was restricted to fewer regions than we have found in otherwise healthy smokers. Critically, we found that smokers with schizophrenia who have lower β_2*-nAChR availability report more negative symptoms and perform worse on tests of executive control. This is important because there are medications available to treat the positive symptoms of schizophrenia (e.g., hallucinations and delusions), but there are no effective medications for the negative symptoms (e.g., anhedonia and flat affect) that significantly impair quality of life. Overall, these data suggest that the cholinergic system may be an excellent target for innovative pharmacological tools to treat some symptoms of schizophrenia and comorbid tobacco addiction.

5 Conclusion

Receptor imaging has led to many insights into tobacco smoking addiction in the living human brain. These findings should help shape our treatment of tobacco smoking. For example, we know that the dopaminergic signature in response to tobacco smoking is short-lived but is associated with significant reductions in craving. We also know that the smoking-induced upregulation of β_2*-nAChRs takes approximately 6–8 weeks to return to control levels. Thus the time course of withdrawal that may contribute to difficulties with quitting smoking has two components that need to be treated. The first is the acute withdrawal (hours, overnight) phase, during which intense craving can be relieved by smoking a cigarette and eliciting dopamine release. The second is a much more prolonged withdrawal phase, during which, there is a normalization of the cholinergic system that may underlie longer-term adaptations in areas such as cognitive functioning and cue-reactivity. Finally, imaging studies should be used to guide drug development for tobacco smokers who are more treatment resistant, such as individuals with psychiatric comorbidity.

Acknowledgments *Funding* K02 DA031750 (Cosgrove) and K01 MH092681 (Esterlis).

References

Abreu-Villaca Y, Seidler FJ, Qiao D, Tate CA, Cousins MM, Thillai I, Slotkin TA (2003) Short-term adolescent nicotine exposure has immediate and persistent effects on cholinergic systems: critical periods, patterns of exposure, dose thresholds. Neuropsychopharmacology 28:1935–1949

Barrett SP, Boileau I, Okker J, Pihl RO, Dagher A (2004) The hedonic response to cigarette smoking is proportional to dopamine release in the human striatum as measured by positron emission tomography and [11C]raclopride. Synapse 54:65–71

Benwell M, Balfour D, Anderson J (1988) Evidence that tobacco smoking increases the density of (−)-[^{3}H]nicotine binding site in human brain. J Neurochem 50:1243–1247

Breese C, Marks M, Logel J, Adams C, Sullivan B, Collins A, Leonard S (1997) Effect of smoking history on [^{3}H]nicotine binding in human postmortem brain. J Pharmacol Exp Ther 282:7–13

Breese C, Lee M, Adams C, Sullivan B, Logel J, Gillen K, Marks M, Collins A, Leonard S (2000) Abnormal regulation of high affinity nicotinic receptors in subjects with schizophrenia. Neuropsychopharmacology 23:351–364

Brody AL, Olmstead RE, London ED (2004) Smoking-induced ventral striatum dopamine release. Am J Psychiatry 161:1211–1218

Brody AL, Mandelkern MA, London ED, Olmstead RE, Farahi J, Scheibal D, Jou J, Allen V, Tiongson E, Chefer SI, Koren AO, Mukhin AG (2006a) Cigarette smoking saturates brain alpha4 beta2 nicotinic acetylcholine receptors. Arch Gen Psychiatry 63:907–915

Brody AL, Mandelkern MA, Olmstead RE, Scheibal D, Hahn E, Shiraga S, Zamora-Paja E, Farahi J, Saxena S, London ED, McCracken JT (2006b) Gene variants of brain dopamine pathways and smoking-induced dopamine release in the ventral caudate/nucleus accumbens. Arch Gen Psychiatry 63:808–816

Brody AL, Mandelkern MA, Costello MR, Abrams AL, Scheibal D, Farahi J, London ED, Olmstead RE, Rose JE, Mukhin AG (2009) Brain nicotinic acetylcholine receptor occupancy: effect of smoking a denicotinized cigarette. Int J Neuropsychopharmacol 12:305–316

Brody AL, London ED, Olmstead RE, Allen-Martinez Z, Shulenberger S, Costello MR, Abrams AL, Scheibal D, Farahi J, Shoptaw S, Mandelkern MA (2010) Smoking-induced change in intrasynaptic dopamine concentration: effect of treatment for tobacco dependence. Psychiatry Res 183:218–224

Brody AL, Mandelkern MA, London ED, Khan A, Kozman D, Costello MR, Vellios EE, Archie MM, Bascom R, Mukhin AG (2012) Effect of secondhand smoke on occupancy of nicotinic acetylcholine receptors in brain. Arch Gen Psychiatry 68:953–960

Brody AL, Mukhin AG, La Charite J, Ta K, Farahi J, Sugar CA, Mamoun MS, Vellios E, Archie M, Kozman M, Phuong J, Arlorio F, Mandelkern MA (2013a) Up-regulation of nicotinic acetylcholine receptors in menthol cigarette smokers. Int J Neuropsychopharmacol 16:957–966

Brody AL, Mukhin AG, Stephanie S, Mamoun MS, Kozman M, Phuong J, Neary M, Luu T, Mandelkern MA (2013b) Treatment for tobacco dependence: effect on brain nicotinic acetylcholine receptor density. Neuropsychopharmacology 38:1548–1556

Brunzell DH, Boschen KE, Hendrick ES, Beardsley PM, McIntosh JM (2010) Alpha-conotoxin MII-sensitive nicotinic acetylcholine receptors in the nucleus accumbens shell regulate progressive ratio responding maintained by nicotine. Neuropsychopharmacology 35:665–673

Collins A, Romm E, Wehner J (1990) Dissociation of the apparent relationship between nicotine tolerance and up-regulation of nicotinic receptors. Brain Res Bull 25:373–379

Cosgrove KP, Batis J, Bois F, Maciejewski PK, Esterlis I, Kloczynski T, Stiklus S, Krishnan-Sarin S, O'Malley S, Perry E, Tamagnan G, Seibyl JP, Staley JK (2009) beta2-Nicotinic acetylcholine receptor availability during acute and prolonged abstinence from tobacco smoking. Arch Gen Psychiatry 66:666–676

Cosgrove KP, Esterlis I, McKee S, Bois F, Alagille D, Tamagnan GD, Seibyl JP, Krishnan-Sarin S, Staley JK (2010) Beta2* nicotinic acetylcholine receptors modulate pain sensitivity in acutely abstinent tobacco smokers. Nicotine Tob Res 12:535–539

Cosgrove KP, Esterlis I, McKee SA, Bois F, Seibyl JP, Mazure CM, Krishnan-Sarin S, Staley JK, Picciotto MR, O'Malley SS (2012) Sex differences in availability of beta2*-nicotinic acetylcholine receptors in recently abstinent tobacco smokers. Arch Gen Psychiatry 69:418–427

D'Souza D, Esterlis I, Carbuto M, Krasenics M, Seibyl J, Bois F, Pittman B, Ranganathan M, Cosgrove K, Staley J (2012) Lower β2*-nicotinic acetylcholine receptor availability in smokers with schizophrenia. Am J Psychiatry 169:326–334

Di Chiara G, Imperato A (1988) Drugs abused by humans preferentially increase synaptic dopamine concentrations in the mesolimbic system of freely moving rats. Proc Natl Acad Sci USA 85:5274–5278

Epping-Jordan M, Picciotto M, Changeux J, Pich EM (1999) Assessment of nicotinic acetylcholine receptor subunit contributions to nicotine self-administration in mutant mice. Psychopharmacology 147:25–26

Esterlis I, Mitsis EM, Batis JC, Bois F, Picciotto MR, Stiklus SM, Kloczynski T, Perry E, Seibyl JP, McKee S, Staley JK, Cosgrove KP (2011) Brain beta2*-nicotinic acetylcholine receptor occupancy after use of a nicotine inhaler. Int J Neuropsychopharmacol 14:389–398

Esterlis I, Hannestad JO, Perkins E, Bois F, D'Souza DC, Tyndale RF, Seibyl JP, Hatsukami DM, Cosgrove KP, O'Malley SS (2013) Effect of a nicotine vaccine on nicotine binding to beta2*-nicotinic acetylcholine receptors in vivo in human tobacco smokers. Am J Psychiatry 170:399–407

Girgis RR, Xu X, Miyake N, Easwaramoorthy B, Gunn RN, Rabiner EA, Abi-Dargham A, Slifstein M (2011) In vivo binding of antipsychotics to D(3) and D(2) receptors: a PET study in baboons with [(11)C]-(+)-PHNO. Neuropsychopharmacology 36:887–895

Imperato A, Mulas A, DiChiara G (1986) Nicotine preferentially stimulates dopamine release n the limbic system of freely moving rats. Eur J Pharmacol 132:337–338

Kassiou M, Eberl S, Meikle S, Birrell A, Constable C, Fulham M, Wong D, Musachio J (2001) In vivo imaging of nicotinic receptor upregulation following chronic (−)-nicotine treatment in baboon using SPECT. Nuc Med Biol 28:165–175

Kim SJ, Sullivan JM, Wang S, Cosgrove KP, Morris ED (2014) Voxelwise lp-ntPET for detecting localized, transient dopamine release of unknown timing: sensitivity analysis and application to cigarette smoking in the PET scanner. Hum Brain Mapp 35:4876–4891

Koranda JL, Cone JJ, McGehee DS, Roitman MF, Beeler JA, Zhuang X (2013) Nicotinic receptors regulate the dynamic range of dopamine release in vivo. J Neurophysiol 111:103–111

Koylu E, Demirgoren S, London E, Pogun S (1997) Sex difference in up-regulation of nicotinic acetylcholine receptors in rat brain. Life Sci 61:PL185–PL190

Laruelle M (2000) Imaging synaptic neurotransmission with in vivo binding competition techniques: a critical review. J Cereb Blood Flow Metab 20:423–451

Le Foll B, Guranda M, Wilson AA, Houle S, Rusjan PM, Wing VC, Zawertailo L, Busto U, Selby P, Brody AL, George TP, Boileau I (2013) Elevation of dopamine induced by cigarette smoking: novel insights from a [(11)C]-(+)-PHNO PET study in humans. Neuropsychopharmacology 39:415–424

Mamede M, Ishizu K, Ueda M, Mukai T, Iida Y, Kawashima H, Fukuyama H, Togashi K, Saji H (2007) Temporal change in human nicotinic acetylcholine receptor after smoking cessation: 5IA SPECT study. J Nucl Med 48:1829–1835

Marks MJ, Pauly JR, Gross SD, Deneris ES, Hermans-Borgmeyer I, Heinemann SF, Collins AC (1992) Nicotine binding and nicotinic receptor subunit RNA after chronic nicotine treatment. J Neurosci 12:2765–2784

Marubio LM, del Mar Arroyo-Jimenez M, Cordero-Erausquin M, Lena C, Le Novere N, de Kerchove d'Exaerde A, Huchet M, Damaj MI, Changeux JP (1999) Reduced antinociception in mice lacking neuronal nicotinic receptor subunits. Nature 398:805–810

Mochizuki T, Villemagne V, Scheffel U, Dannals R, Finley P, Zhan Y, Wagner H, Musachio J (1998) Nicotine induced up-regulation of nicotinic receptors in CD-1 mice demonstrated with an in vivo radiotracer: gender differences. Synapse 30:116–118

Montgomery AJ, Lingford-Hughes AR, Egerton A, Nutt DJ, Grasby PM (2007) The effect of nicotine on striatal dopamine release in man: a [11C]raclopride PET study. Synapse 61:637–645

Morris ED, Kim SJ, Sullivan JM, Wang S, Normandin MD, Constantinescu CC, Cosgrove KP (2013) Creating dynamic images of short-lived dopamine fluctuations with lp-ntPET: dopamine movies of cigarette smoking. J Vis Exp 78

Mukhin AG, Kimes AS, Chefer SI, Matochik JA, Contoreggi CS, Horti AG, Vaupel DB, Pavlova O, Stein EA (2008) Greater nicotinic acetylcholine receptor density in smokers than in nonsmokers: a PET study with 2-18F-FA-85380. J Nucl Med 49:1628–1635

Perkins KA (2009) Sex differences in nicotine reinforcement and reward: influences on the persistence of tobacco smoking. Nebr Symp Motiv 55:143–169

Perkins KA, Scott J (2008) Sex differences in long-term smoking cessation rates due to nicotine patch. Nicotine Tob Res 10:1245–1250

Perkins KA, Donny E, Caggiula AR (1999) Sex differences in nicotine effects and self-administration: review of human and animal evidence. Nicotine Tob Res 1:301–315

Picciotto M, Zoli M, Rimondin R, Lena C, Marubio L, Pich E, Fuxe K, Changeux J (1998) Acetycholine receptors containing the beta2 subunit are involved in the reinforcing properties of nicotine. Nature 391:173–177

Pietila K, Lahde T, Attila M, Ahtee L, Nordberg A (1998) Regulation of nicotinic receptors in the brain of mice withdrawn from chronic oral nicotine treatment. Naunyn-Schmiedeberg's Arch Pharmacol 357:176–182

Rose JE, Mukhin AG, Lokitz SJ, Turkington TG, Herskovic J, Behm FM, Garg S, Garg PK (2010) Kinetics of brain nicotine accumulation in dependent and nondependent smokers assessed with PET and cigarettes containing 11C-nicotine. Proc Natl Acad Sci USA 107:5190–5195

Scott DJ, Domino EF, Heitzeg MM, Koeppe RA, Ni L, Guthrie S, Zubieta JK (2007) Smoking modulation of mu-opioid and dopamine D2 receptor-mediated neurotransmission in humans. Neuropsychopharmacology 32:450–457

Srinivasan R, Pantoja R, Moss FJ, Mackey ED, Son CD, Miwa J, Lester HA (2010) Nicotine up-regulates alpha4beta2 nicotinic receptors and ER exit sites via stoichiometry-dependent chaperoning. J Gen Physiol 137:59–79

Staley JK, Krishnan-Sarin S, Cosgrove KP, Krantzler E, Frohlich E, Perry E, Dubin JA, Estok K, Brenner E, Baldwin RM, Tamagnan GD, Seibyl JP, Jatlow P, Picciotto MR, London ED, O'Malley S, van Dyck CH (2006) Human tobacco smokers in early abstinence have higher levels of beta2* nicotinic acetylcholine receptors than nonsmokers. J Neurosci 26:8707–8714

Sullivan JM, Kim SJ, Cosgrove KP, Morris ED (2013) Limitations of SRTM, Logan graphical method, and equilibrium analysis for measuring transient dopamine release with [(11)C] raclopride PET. Am J Nucl Med Mol Imaging 3:247–260

Takahashi H, Fujimura Y, Hayashi M, Takano H, Kato M, Okubo Y, Kanno I, Ito H, Suhara T (2008) Enhanced dopamine release by nicotine in cigarette smokers: a double-blind, randomized, placebo-controlled pilot study. Int J Neuropsychopharmacol 11:413–417

Tritto T, McCallum SE, Waddle SA, Hutton SR, Paylor R, Collins AC, Marks MJ (2004) Null mutant analysis of responses to nicotine: deletion of beta2 nicotinic acetylcholine receptor subunit but not alpha7 subunit reduces sensitivity to nicotine-induced locomotor depression and hypothermia. Nicotine Tob Res 6:145–158

Tziortzi AC, Searle GE, Tzimopoulou S, Salinas C, Beaver JD, Jenkinson M, Laruelle M, Rabiner EA, Gunn RN (2011) Imaging dopamine receptors in humans with [11C]-(+)-PHNO: dissection of D3 signal and anatomy. Neuroimage 54:264–277

Behavioral Mechanisms Underlying Nicotine Reinforcement

Laura E. Rupprecht, Tracy T. Smith, Rachel L. Schassburger, Deanne M. Buffalari, Alan F. Sved and Eric C. Donny

Abstract Cigarette smoking is the leading cause of preventable deaths worldwide, and nicotine, the primary psychoactive constituent in tobacco, drives sustained use. The behavioral actions of nicotine are complex and extend well beyond the actions of the drug as a primary reinforcer. Stimuli that are consistently paired with nicotine can, through associative learning, take on reinforcing properties as conditioned stimuli. These conditioned stimuli can then impact the rate and probability of behavior and even function as conditioning reinforcers that maintain behavior in the absence of nicotine. Nicotine can also act as a conditioned stimulus (CS), predicting the delivery of other reinforcers, which may allow nicotine to acquire value as a conditioned reinforcer. These associative effects, establishing non-nicotine stimuli as conditioned stimuli with discriminative stimulus and conditioned reinforcing properties as well as establishing nicotine as a CS, are predicted by basic conditioning principles. However, nicotine can also act non-associatively. Nicotine directly enhances the reinforcing efficacy of other reinforcing stimuli in the environment, an effect that does not require a temporal or predictive relationship between nicotine and either the stimulus or the behavior. Hence, the reinforcing actions of nicotine stem both from the primary reinforcing actions of the drug (and the subsequent associative learning effects) as well as the reinforcement enhancement action of nicotine which is non-associative in nature. Gaining a better understanding of how nicotine impacts behavior will allow for maximally effective tobacco control efforts aimed at reducing the harm associated with tobacco use by reducing and/or treating its addictiveness.

Keywords Reinforcement · Reward · Operant · Self-administration · Nicotine · Conditioning

---

L.E. Rupprecht · R.L. Schassburger · D.M. Buffalari · A.F. Sved
Department of Neuroscience, University of Pittsburgh, Pittsburgh, PA 15260, USA

T.T. Smith · A.F. Sved · E.C. Donny (✉)
Department of Psychology, University of Pittsburgh, 3137 Sennott Square, 210 S. Bouquet Street, Pittsburgh, PA 15260, USA
e-mail: edonny@pitt.edu

D.J.K. Balfour and M.R. Munafò (eds.), *The Neuropharmacology of Nicotine Dependence*, Current Topics in Behavioral Neurosciences 24,
DOI 10.1007/978-3-319-13482-6_2

Contents

1 Introduction

Cigarette smoking is the leading cause of preventable morbidity and mortality worldwide, resulting in about 4.9 million deaths per year. It is widely accepted that nicotine is the primary psychoactive and reinforcing component of tobacco that produces the addictive state underlying sustained use of cigarettes and other tobacco products (Stolerman and Jarvis 1995; USDHHS 1988).

Despite the clinical observations that tobacco products are quite addictive and success rates for quitting tobacco use are low (USDHHS 1988, 2012), experimental evidence suggests the primary reinforcing properties of nicotine, by itself, are weak. Indeed, when considering the primary reinforcing properties of nicotine compared to other drugs of abuse, the addictive power of tobacco products is surprising. However, the reinforcing properties of nicotine are much more complex than simple primary reinforcement from the drug. In addition to nicotine acting as a primary reinforcer, other stimuli that are consistently paired with nicotine can, through associative learning, take on reinforcing properties as conditioned stimuli. These conditioned stimuli can then impact the rate and probability of behavior and even function as conditioning reinforcers that maintain behavior in the absence of nicotine.

Nicotine can also act as a conditioned stimulus (CS), predicting the delivery of other reinforcers, which may allow nicotine to acquire value as a conditioned reinforcer. These associative effects, establishing non-nicotine stimuli as conditioned stimuli with discriminative stimulus and conditioned reinforcing properties as well as establishing nicotine as a CS, are predicted by basic conditioning principles. However, nicotine can also act non-associatively. Nicotine directly enhances the reinforcing efficacy of other reinforcing stimuli in the environment, an effect that does not require a temporal or predictive relationship between nicotine and either the stimulus or the behavior. Hence, the reinforcing actions of nicotine stem both from the primary reinforcing actions of the drug (and the subsequent associative learning effects) as well as the reinforcement enhancement action of nicotine which is non-associative in nature. Together, these two actions constitute what has been referred to as the "dual-reinforcement" model of nicotine reinforcement (Caggiula et al. 2009; Chaudhri et al. 2006b; Donny et al. 2003). Nicotine as a CS/reinforcer has received less attention, but potentially functions as a third mechanism underlying nicotine reinforcement (Bevins 2009). These actions serve as the basis for this chapter.

2 Nicotine Self-administration

The gold standard for studying the reinforcing properties of drugs in experimental animals is self-administration in which animals need to perform a behavioral task (e.g., press a lever, poke their nose into a hole) to obtain the drug. Hence, the focus of the chapter is on self-administration studies. Data from other procedures such as conditioned place preference (CPP), in which animals choose between an environment previously paired with the drug and one never paired with the drug, are incorporated when they provide additional insight. Most nicotine self-administration studies have used rats (Corrigall and Coen 1989; Donny et al. 1995), although self-administration has been demonstrated across a wide range of species including humans, non-human primates, dogs, and, more recently, mice (Corrigall 1999; Fowler and Kenny 2011; Goldberg et al. 1983; Le Foll and Goldberg 2005; Matta et al. 2007; Stolerman 1999).

However, it is important to appreciate that depending on the details of the methodological approach, multiple actions of nicotine might be occurring together and underlie the observed behavior. Historically, nicotine self-administration procedures in rats have utilized a protocol in which nicotine delivery is paired with a stimulus such as a cue light, chamber light, or tone. That stimulus might function as a primary reinforcer and/or acquire conditioned reinforcing properties after repeated association with nicotine infusion delivery. The reinforcing effects of that stimulus might then also be affected by the non-associative effects of nicotine. Thus, self-administration procedures often do not distinguish between primary reinforcement, reinforcement enhancement, and associative effects that may emerge over repeated experimental sessions. When considering data from nicotine self-administration studies, it is essential to understand the details of the methodological approach and be mindful of the different actions of nicotine that may underlie the observed behavior. This complexity is arguably an appropriate model for tobacco use in people in which the interaction between nicotine and concurrent stimuli is reality and these multiple reinforcing actions of nicotine are integrated to support behavior.

The reinforcing effects of nicotine are affected by a number of important moderating variables. For example, the duration and route of administration of nicotine delivery and the sex and age of subjects impact the reinforcing properties of nicotine and are highly relevant when considering the reinforcing properties of tobacco products. Furthermore, in the context of tobacco products, other chemical constituents found in tobacco products that would be delivered along with nicotine may interact with nicotine to alter its complex reinforcing properties. Moderating variables are discussed in more detail below.

Although humans typically self-administer nicotine in the context of tobacco products, laboratory studies have shown than people will work for nicotine. This research, typically conducted with infusions of nicotine paired with a novel light stimulus, shows that nicotine supports behavior that is dose dependent and schedule dependent and different from vehicle (Harvey et al. 2004; Henningfield and Goldberg 1983). Smokers also report subjective experiences consistent with the rewarding effects of nicotine (Sofuoglu et al. 2008). Although limited, these human laboratory studies confirm the basic observations from experimental animals, supporting the validity of the preclinical models that form the basis of this chapter.

2.1 Outcome Measures

It is worth noting that while most self-administration studies focus on measuring response or infusion rates during stable periods of self-administration as the key dependent measures, there are several other important measures that may reflect the reinforcing properties of nicotine. In particular, the rate at which rats acquire self-administration likely reflects the reinforcing actions of nicotine (i.e., the more the reinforcing, the faster the rats will acquire the behavior). Similarly, the percentage of rats self-administering nicotine should also reflect the reinforcing properties of nicotine under those conditions. Like stable rates of nicotine self-administration that are maintained over multiple sessions, the rate and percentage of rats acquiring is dose related and might be particularly sensitive to changes in the reinforcing properties of nicotine. For example, self-administration on a fixed ratio (FR) 2 schedule for a low dose of nicotine (7.5 μg/kg/infusion) compared with a maximally effective dose (60 μg/kg/infusion) develops more slowly and in a smaller percentage of rats, but the level of self-administration ultimately attained is similar (Smith et al. 2014b). Many researchers also measure the willingness of an animal to work for a reinforcer, in this case infusions of nicotine, by increasing the number of responses required for an infusion within a session [progressive ratio (PR)] (Donny et al. 1999). Other measures such as the latency to start responding or earn an infusion, responding during a time out period, or responding despite an additional negative consequence have received relatively little attention but might provide additional insight into the factors controlling behavior (e.g., latency to first infusion may be a reflection of the influence of the context on behavior).

2.2 *Route, Dose, and Rate of Nicotine*

The pharmacological actions of nicotine are mediated through its interactions with nicotinic acetylcholine receptors (nAChR), which are located in many sites throughout the brain and rest of the body (Leslie et al. 2013). Multiple subtypes of nAChR comprised of many subunit compositions have differing affinities for nicotine and kinetics (Picciotto et al. 2001). Thus, it is not surprising that the reinforcing actions of nicotine are dependent upon dose, route of administration, and temporal aspects of delivery (Le Foll and Goldberg 2005).

Nicotine self-administration protocols predominantly use methods that deliver infusions intravenously over a brief period of 1–5 s (Matta et al. 2007). The use of short-duration intravenous (iv) infusions is based on the understanding that drugs that rapidly reach the brain are considered more reinforcing and have increased abuse liability (Benowitz 1990; Samaha and Robinson 2005), and the assumption that nicotine inhaled from cigarette smoke is rapidly absorbed into the pulmonary circulation and thus reaches the brain 5–10 s following inhalation (Russell and Feyerabend 1978). Additional puffs of cigarette smoke provide additional pulses of nicotine, superimposed on an increasing blood level of nicotine (Rose et al. 1999). Importantly, rapid discrete (iv) infusions of nicotine support self-administration (Corrigall and Coen 1989; Donny et al. 1995; Matta et al. 2007). In self-administration procedures that use cues (e.g., a cue light located above the active lever or other operandum associated with (iv) nicotine delivery), rats reliably self-administer nicotine in the 10–60 μg/kg/infusion range (Chaudhri et al. 2007; Donny et al. 1995, 1998, 1999). Several studies directly examining the impact of prolonging infusion duration on self-administration of nicotine (Wakasa et al. 1995; Wing and Shoaib 2013) determined that short-duration infusion delivery supported greater responding, as might be expected from other abuse liability research (Ator and Griffiths 2003; Busto and Sellers 1986; Farre and Cami 1991; McColl and Sellers 2006; Samaha and Robinson 2005).

On the other hand, it has recently been argued that the rapid (iv) infusion of nicotine (<5 s) typically used in rat self-administration studies, coupled with single infusion doses that are on par with the quantity of nicotine delivered in an entire cigarette (i.e., 10–30 μg/kg/cigarette for a 70 kg individual), does not accurately represent the increase in plasma nicotine concentration of a person smoking a cigarette (Sorge and Clarke 2009; Rose et al. 1999). To model the dose of a single cigarette puff and the seconds-to-minutes long rise in arterial nicotine concentration, smaller doses delivered over a longer infusion duration may be required. "Puff-sized" doses modeling that delivered in a single puff of a cigarette might be closer to 1–10 μg/kg nicotine, and the (iv) infusion duration might be several tens of seconds to better model tissue nicotine delivery from smoking cigarettes. Sorge and Clarke (2009) performed a series of experiments designed to more closely mimic nicotine delivery in human smokers and determine optimal dose/duration combinations chosen by rats in a concurrent access protocol. First, to test the reinforcing value of nicotine (15 μg/kg/infusion) delivered over different lengths of time, adult male rats self-administered infusions delivered over 3, 30, 60, or 120 s in a cued protocol. On a FR 1 schedule of reinforcement, there was no effect of infusion duration on infusions earned; however, on an FR 5 schedule, rats self-administered equal numbers of infusions of 3 and 30 s lengths, which were significantly greater than more prolonged infusion durations. Rats given the opportunity to respond on different levers to deliver 3- or 30-s infusions of nicotine (15 μg/kg) selected the 30-s infusion duration significantly more than the 3-s duration (10 vs. 6 infusions during 2-h sessions). Analysis of the dose–response effects of nicotine using this prolonged 30-s infusion duration determined that on an FR 1 schedule, doses between 3 and 30 μg/kg supported self-administration, and all doses above 3 μg/kg supported self-administration on an FR 5 schedule of reinforcement. The authors concluded that rats prefer slow infusions to fast when given a choice and that rats will self-administer a wider range of nicotine doses, including "puff-sized" doses, when the infusion duration is prolonged to this 30-s range.

Other data, however, are inconsistent with this observation. The low doses shown to support behavior with prolonged infusions (i.e., 3 and 10 μg/kg/infusion) have also supported robust self-administration with 1-s infusions in cued paradigms (Bardo et al. 1999; Donny et al. 1995; Smith et al. 2013, 2014b). Furthermore, unpublished data from our laboratory comparing infusions of nicotine delivered over 1 or 20 s found little difference in responding for both 3.75 and 10 μg/kg/infusion nicotine. Groups exhibited a difference in responding for 30 μg/kg nicotine, with animals receiving the rapid 1-s infusion earning nearly double the infusions earned by the group receiving slow 20-s infusions. Other studies have produced mixed results with moderate increases in infusion duration of nicotine finding either no evidence of self-administration using 6-s infusions or even a moderate but significant increase in infusions earned by mice receiving 3-s compared to 1-s nicotine infusions (Belluzzi et al. 2005; Fowler and Kenny 2011).

What is clear from the published literature is that there is no consensus and considerable variability, regarding the impact of infusions in the range of 1 s ("rapid" infusion), which has been standard in most nicotine self-administration studies, to tens of seconds ("slow" infusion). Discussion of the rate of (iv) delivery should not simply focus on what rate supports the most robust self-administration. What infusion profile best models nicotine delivery in people depends on the tobacco product or nicotine delivery system (e.g., electronic cigarette or nicotine patch) being considered. This issue will become increasingly important as new nicotine delivery devices are developed and their abuse liability is evaluated. Additionally, simply comparing rates of nicotine self-administration may miss important differences in underlying mechanism. For example, Sorge and Clarke (2009) have reported the intriguing observation that dopamine antagonists differentially impact nicotine self-administration depending upon whether nicotine is delivered via "rapid" or "slow" infusions. Likewise, other moderating variables (e.g., cues) may be more or less important depending on the rate of nicotine delivery. Indeed, whereas the primary reinforcing actions of nicotine, along with the associative actions derived from them, require discrete infusions of nicotine contingent on behavior, the non-associative reinforcement enhancement actions of nicotine are observed with slow, even systemic, administration, as well as rapid (iv) infusions (Caggiula et al. 2009).

2.3 Food Restriction

Most self-administration studies of addictive drugs, including nicotine, are conducted on mildly food-restricted rats, with food limited to approximately 80 % of the amount of food consumed with unlimited access. This food restriction maintains weight gain, though at a rate below what is observed with unlimited access to food. Experimental animal models of drug abuse use food restriction for a variety of reasons. Constraining growth might prolong patency of the catheter. It might also be argued that animals are not normally exposed to easy access to unlimited food sources and that unlimited access results in overweight and unhealthy animals (Abelson 1995; Speakman and Mitchel 2011). Most importantly, food restriction leads to higher levels of responding in operant procedures and contributes to the overall motivational state of the animal (Lang et al. 1977). Food restriction typically results in more robust self-administration and self-administration of low doses of nicotine that is not observed without food restriction (Corrigall and Coen 1989; Donny et al. 1998; Singer et al. 1978). This increase in self-administration may simply result from the animals being more likely to explore their environment and respond on operanda, as chronic calorie restriction leads to an increase in physical activity prior to food availability (Duffy et al. 1990; Russell et al. 1987). Interestingly, there is evidence that calorie restriction also increases cigarette use in human smokers (Cheskin et al. 2005). It is important to emphasize, however, that food restriction is not required for rats to acquire self-administration behavior for nicotine (Donny et al. 1998; Peartree et al. 2012; Yan et al. 2012). Whatever the explanation for why moderate food restriction enhances nicotine self-administration (and responding in other operant procedures as well), the imposition of this "motivational state" is an important variable in nicotine self-administration procedures.

2.4 Session Length

The vast majority of nicotine self-administration studies rely on daily (or 5 day per week) sessions lasting an hour or a few hours. Other studies allow more continual self-administration access, allowing rats to respond for nicotine infusions for 22 or 23 h/day with a short break for cleaning the operant chambers. Rats readily self-administer nicotine in both limited and extended access procedures, and importantly, the dose–response relationship for self-administration is similar (Matta et al. 2007). Extended access procedures result in greater daily nicotine intake, although the rate of infusions per hour is reduced. Given the more continuous nature of extended access, animals are more likely to develop nicotine dependence as a results of nicotine exposure that is maintained for a prolonged period each day and may undergo withdrawal if access to nicotine is terminated or pharmacologically precipitated (O'Dell et al. 2007), or even just with significant gaps in self-administration. Consequently, dependence may result in another reinforcing action supporting additional behavior that is not observed in the limited access procedures, negative reinforcement related to withdrawal suppression. Interestingly, Cohen et al. (2012) observed that if periods of extended access are spaced with one or two days of no access, the rates of self-administration are substantially greater compared to both extended access without those breaks and limited access sessions with similar days of no access. In contrast, others have argued that limited and extended access paradigms produce comparable levels of dependence (Paterson and Markou 2004). Despite these potential differences, most studies continue to use limited access because of practical issues and limited access protocols are well suited to examine the reinforcing properties of nicotine in the absence of issues associated with withdrawal.

2.5 Sex and Age

The sex of a subject may also moderate (iv) nicotine self-administration, an important issue to consider given that there are sex differences in tobacco use. Although men are more likely than women to smoke, data from human studies and national surveys of smoking behavior suggest that sex differences are complex and could differentially impact susceptibility to initiate tobacco product use, the progression to dependence, and difficulty with successful cessation (Benowitz and Hatsukami 1998; Brady and Randall 1999; Carroll et al. 2004; Kim and Fendrich 2002; Lynch et al. 2002; Lynch 2006, 2009). These differences may be related to differences in pharmacokinetics and metabolism (Jensvold et al. 1996; Kyerematen et al. 1988), brain development, physiology and function (Berchtold et al. 2008; Chambers et al. 2003), and circulating reproductive hormones (Becker and Hu 2008; Breslau and Peterson 1996; O'Hara et al. 1989; Perkins et al. 1999). Women have also been shown to be more reactive to nicotine-associated cues (Perkins et al. 1999, 2001) and stress, which may lead to a greater propensity to relapse (Perkins et al. 2013; Schnoll et al. 2007; Xu et al. 2008).

Sex differences in tobacco use can be recapitulated in rat models of nicotine self-administration suggesting a biological basis. Adult female rats have been shown to respond more for nicotine, although they also demonstrated increased responding on an inactive lever (Chaudhri et al. 2005). In other studies using nicotine paired with a visual stimulus (VS) (discussed more below), male and female rats acquired nicotine self-administration at a comparable rate and to similar asymptotic levels within a standard range of nicotine doses (30–90 μg/kg/infusion), but females acquired self-administration faster and reached higher break points on a PR schedule of reinforcement at lower doses of nicotine (Donny et al. 2000; Lanza et al. 2004). In a 23-h extended access test, adult female rats had higher rates of responding for a large dose (60 μg/kg/infusion) than adult males rats (Grebenstein et al. 2013). Finally, females may experience greater withdrawal from nicotine, particularly those related to anxiety/stress (for review, see O'Dell and Torres 2014). However, to our knowledge, sex differences related to negative reinforcement have not been evaluated.

Similarly, the age of a subject may also influence nicotine reinforcement. Approximately 90 % of adult daily smokers initiated use prior to the age of 18, and nearly all adult smokers began prior to the age of 25 (CDC 2012; USDHHS 2012). National surveys of high school students report that 23 % of students are current users of tobacco products (use on at least 1 of the last 30 days) (CDC 2010). Earlier initiation of smoking increases the likelihood that someone will become a heavy smoker, be more dependent, and have greater difficulty quitting (USDHHS 2012). These epidemiological data suggest that adolescence represents a period of vulnerability to nicotine use and dependence. Adolescents are biologically driven to seek novelty and risk (Spear 2000) and are also more reactive to stress and unable to modulate their response in a productive way (Chambers et al. 2003). Although the preclinical literature examining developmental differences in nicotine use and reinforcement is limited and varied in approach, collectively, the findings suggest that there are differences in the reinforcement of nicotine between adolescents and adults.

Adolescent rats have been shown to self-administer nicotine at rates compared to adults across a range of doses (Chen et al. 2007; Levin et al. 2007, 2011; Natividad et al. 2013). In a direct comparison of adolescent to adult females, adolescent females self-administer significantly more nicotine than adults (Levin et al. 2003). A study testing nicotine self-administration in male and female adolescents found that both sexes responding similarly for nicotine infusions compared to adult rats (Chen et al. 2007); however, female adolescent rats acquired nicotine self-administration more quickly and reached higher rates of responding than adult rats. Lynch (2009) evaluated differences in self-administration of nicotine between male and female adolescents in a long-access paradigm. Again, female adolescents displayed the most robust self-administration behavior under a FR 1 and a PR schedule of reinforcement (Lynch 2009).

Another issue that comes up in relation to the adolescent period is whether exposure to nicotine during adolescence alters nicotine self-administration in adulthood. Rats exposed to experimenter-administered nicotine as adolescents showed increased nicotine self-administration (Adriani et al. 2003; Natividad et al. 2013) as adults compared to animals only exposed to nicotine during adulthood. These data in rodents support the epidemiological data and hypothesis that early exposure to nicotine may heighten the reinforcing actions of nicotine in adults. However, studies from Shram et al. (2008a, b) suggest that adolescence is not a period of enhanced sensitivity to the reinforcing actions of nicotine, noting the lack of a difference in nicotine self-administration between adult and adolescent rats on FR 1 and FR 2 schedules of reinforcement, and greater responding by adults on an

FR 5 and PR schedule of reinforcement. Further complicating the influence of age are moderating variables such as stress. When treated with yohimbine, an alpha-2 adrenergic receptor antagonist that many laboratories use to model a stress response, adolescent females reached significantly higher break points and earned more nicotine infusions than adolescent males at all doses of nicotine tested (7.5, 15, 30 μg/kg/infusion), suggesting greater reactivity after a stressor-like challenge (Li et al. 2014; Zou et al. 2014). These data in rodents support the epidemiological data that female adolescents may be more sensitive to the reinforcing properties of nicotine, particularly with an additional stressor (McKee et al. 2003).

In CPP protocols, adolescent rats and mice have been shown to develop a preference after a single nicotine treatment (adults do not), exhibit a larger degree of preference for nicotine than adults, including at lower doses of nicotine, and lack of an aversion at high doses of nicotine (Belluzzi et al. 2004; Brielmaie et al. 2007; Kota et al. 2007; Torres et al. 2008; Vastola et al. 2002). Additionally, animals exposed to nicotine as adolescents showed increased CPP as adults (Adriani et al. 2006).

In sum, the preclinical literature predominantly supports the epidemiological data from smokers that nicotine reinforcement may differ between males and females and that adolescence is a period of heightened susceptibility to the reinforcing effects of nicotine. Additional research utilizing animal models is critical for understanding the biological basis of age- and sex-related differences in nicotine reinforcement. Although these studies can be technically challenging (e.g., catheter patency in young animals, estrous cycle variation), they provide an opportunity for experimental manipulations that are not possible in a clinical setting and therefore can contribute to a better understanding of the causal relationship between development, sex, and nicotine reinforcement.

2.6 Other Compounds in Cigarette Smoke

The reinforcing properties of nicotine can be modified by other chemicals in the tobacco product. In cigarettes, nicotine is typically taken along with more than 8,000 other chemicals (CDC 2010). Research on several of the compounds in tobacco smoke have suggested that they may be psychoactive and have abuse potential by themselves and potentiate the abuse liability of nicotine (Hoffman and Evans 2013). Non-nicotine compounds in cigarette smoke that may have reinforcing value or contribute to the reinforcing actions of nicotine include, but are not limited to, acetaldehyde, minor alkaloids (e.g., nornicotine, myosmine, cotinine, anabasine, anatabine), and ß-carbolines (e.g., harman and norharman). Investigating these smoke constituents, alone and in combination with nicotine, is critical for a better understanding of the reinforcing properties of cigarettes. However, the best methodology for studying the reinforcing potential of these constituents is complicated. For example, given that most of these constituents exist in cigarette smoke at concentrations much lower than nicotine, what doses are appropriate for investigation? Should constituents be investigated in isolation or combined?

Acetaldehyde, one smoke constituent, has received a fair amount attention. It is one of several aldehydes present in tobacco smoke (Houlgate et al. 1989; Xie et al. 2009), resulting from the combustion of polysaccharides, as well as being added in the manufacture of commercial cigarettes. Acetaldehyde is also the major metabolite of ethanol and has previously been a focus of alcohol-abuse liability research (Correa et al. 2012; Deng and Deitrich 2008). Nicotine and alcohol are often consumed together, adding to the importance of studying the interaction between nicotine and acetaldehyde. Acetaldehyde by itself acts a reinforcer and has been shown to be self-administered orally (Peana et al. 2010, 2011), intravenously (Myers et al. 1982, 1984a, b; Takayama and Uyeno, 1985), and directly into the brain (Amit et al. 1977; Brown et al. 1979; Rodd-Hendricks et al. 2002) by rats. Animals also prefer chambers paired with acetaldehyde in CPP experiments (Melis et al. 2007; Quertemont and De Witte 2001; Peana et al. 2008; Smith et al. 1984; Spina et al. 2010). Taken together, these results suggest that acetaldehyde has reinforcing properties.

Importantly, acetaldehyde may also interact with nicotine to potentiate the magnitude of reinforcement. Early studies within the tobacco industry examined whether acetaldehyde might potentiate nicotine self-administration. Denoble and colleagues tested the ability of nicotine and acetaldehyde to be self-administered by rats alone or in combination, across a range of doses (0–16 μg/kg/infusion for both drugs) with the goal of isolating dose combinations that would produce the highest levels of reinforcement (DeNoble and Mele 1983). Importantly, acetaldehyde supported higher levels of responding than nicotine at equal doses. The combination of nicotine and acetaldehyde enhanced responding for infusions as compared to responding for infusions of either drug alone. The augmented responding for the combination was most robust at low doses of nicotine (2–8 μg/kg/infusion) and at the highest dose of acetaldehyde (16 μg/kg/infusion) tested. The doses of nicotine and acetaldehyde required to produce maximal responding for (iv) infusions were quite different from what is in cigarettes, where acetaldehyde concentrations are approximately half that of what is observed for nicotine. More recently, Belluzzi et al. (2005) reported that adolescent male rats robustly self-administered a mix of nicotine (30 μg/kg/infusion) and acetaldehyde (16 μg/kg/infusion) while not self-administering either substance individually, which better models the nicotine-to-acetaldehyde ratio in cigarettes. Interestingly, this interaction between acetaldehyde and nicotine was not observed in adult rats. However, this study used a somewhat unusual set of conditions (i.e., 5.6 s infusions, ad libitum feeding, FR 1, 3-h sessions, only 5 self-administration sessions) in which nicotine alone is not self-administered to a significant degree. Taken together, these studies highlight the possibility that acetaldehyde administered along with nicotine can increase the reinforcing properties of nicotine, at least under some conditions.

Although nicotine is the primary alkaloid found in tobacco, accounting for roughly 95 % of the alkaloid content, other alkaloids (nornicotine, myosmine, cotinine, anabasine, and anatabine) are also present (Huang and Hsieh 2007). These minor alkaloids are similar in structure to nicotine, and some are metabolites of nicotine (Crooks et al. 1997). A limited body of data suggests that some of these minor alkaloids might have reinforcing properties, but only at doses much higher than or equal to nicotine (Bardo et al. 1999; Caine et al. 2014). In a test of whether rats would self-administer a combination of nornicotine, myosmine, cotinine, anabasine, and anatabine, with doses indexed to their concentration in cigarette smoke relative to nicotine, the alkaloid cocktail did not support self-administration behavior (Clemens et al. 2009). These limited results provide evidence that large doses of some minor alkaloids may have positive reinforcing properties by themselves, but the reinforcing effects of these constituents are likely weak at doses that more closely approximate the levels in tobacco (relative to nicotine). More importantly, this mix of five minor alkaloids appeared to enhance the reinforcing actions of nicotine, especially at lower doses of nicotine (Clemens et al. 2009). Using a cued protocol with 4-s infusions, rats self-administered a solution containing 30 μg/kg/infusion of nicotine along with the minor alkaloids significantly more than just nicotine. The increase in self-administration associated with the co-administration of the minor alkaloids was dependent on the reinforcement schedule (it was observed at FR 5 and PR schedules but not FR 1 or FR 2) and appeared to be larger at smaller doses of nicotine. However, the minor alkaloids co-administered along with nicotine also increased locomotor activity compared to just nicotine and increased inactive responding on the FR 5 schedule to the same extent as it increased active responding, raising questions as to whether this interaction between minor alkaloids and nicotine results from increased reinforcement. Relatedly, acute systemic treatment with anabasine (20 μg/kg), but not anatabine, nornicotine, myosmine, harman, and norharman, increased the number of nicotine infusions (30 μg/kg/infusion) earned by periadolescent female rats (Hall et al. 2014). However, larger doses of anabasine, anatabine, and nornicotine, when administered systemically prior to nicotine self-administration sessions, suppress the number of infusions (Mello et al. 2014; Caine et al. 2014; Hall et al. 2014). Although results are limited and mixed, studies like these emphasize the need for increased attention to the interaction between nicotine and other alkaloids that might naturally be consumed along with nicotine.

An alternative approach to examining whether the additional compounds in cigarettes contribute to the reinforcing properties of nicotine in cigarettes is to evaluate self-administration of an extract produced from tobacco or smoke. Recently, Costello et al. (2014) compared self-administration of an aqueous extract of cigarette smoke to that of pure nicotine in adult male rats. At low concentrations of nicotine (3.75 and 7.5 μg/kg/infusion), self-administration was enhanced by the other components in the extract, but self-administration was not different at the highest dose of nicotine tested (15 μg/kg/infusion). While one interpretation of their data is that the other non-nicotine components in their extract enhanced the reinforcing properties of nicotine, it may instead be that these other chemicals in the extract are themselves reinforcing since there was no test of self-administration of a denicotinized extract. Still, self-administration of nicotine in the extract was attenuated by a nicotinic receptor antagonist, suggesting that effects on the nAChR were important for producing the increase in self-administration.

Recently, significant attention has focused on the potential role of MAO inhibition on nicotine reinforcement. This attention derives from clinical studies demonstrating that both MAO A and MAO B are 30–40 % inhibited in the brains of smokers relative to non-smokers (Berlin et al. 1995; Fowler et al. 1996a, 1996b). Precisely which constituents account for this level of inhibition is unknown. The β-carbolines, harman and norharman, may contribute to both the inhibition of MAO and the impact on nicotine reinforcement. In a study of the impact of monoamine oxidase (MAO) inhibition on nicotine self-administration, norharman was given chronically to inhibit MAO, resulting in a potentiation of self-administration (Guillem et al. 2005). However, the dose of norharman was substantially higher than that actually delivered in tobacco smoke. Since the constituents in cigarette smoke that lead to MAO inhibition are unknown, several studies in rats have attempted to understand the impact of MAO inhibition on nicotine reinforcement using known MAO inhibitors that are not present in tobacco smoke. Using drugs such as tranylcypromine and phenylzine to inhibit MAO, studies have consistently shown increased self-administration of nicotine, especially at low doses of nicotine (Smith et al. 2014a; Villegier et al. 2007a, b). Three points are worth making here. First, large doses of drugs that inhibit MAO appear to increase the reinforcing properties of nicotine (Smith et al. 2014a; Villegier et al. 2007a, b). Second, the interpretation of these data is complicated by the possibility that the effects produced by large doses of MAO inhibitors may be due, at least in part, to actions of these drugs other than MAO inhibition (Loftipour et al. 2011; Villegier et al. 2007a, b). Third, studies published to date have not examined partial inhibition, as seen in smokers.

As the preceding paragraphs make clear, studying which constituents in addition to nicotine that might impact reinforcement may provide important insight into how nicotine, in the form of tobacco products, reinforces behavior. This work is still in its infancy and the challenge of untangling the role of other constituents can be daunting; namely, it is not clear what constituents should demand experimental focus and how to best model the potential interactions between constituents. However, it is also important to recognize that in the absence of nicotine, these data show these other constituents (in the levels found in tobacco products) are not reinforcing. This observation affirms the conclusions reached over the last several decades that nicotine is the primary addictive substance in tobacco. Hence, these other constituents are still best viewed as potential moderators of the effects of nicotine; they do not appear to be sufficient for maintaining tobacco-use behavior.

3 Primary Reinforcing Actions of Nicotine

As described up to this point, rats will self-administer nicotine in operant protocols. However, these are typically cued protocols that do not distinguish among primary reinforcement, reinforcement enhancement (discussed below), and the associative processes that may then result over repeated experimental sessions. If there were no external cues that were associated with nicotine delivery, then only the primary reinforcing action of nicotine would be present to support self-administration behavior. The few studies that have examined nicotine without additional associative cues provide support for nicotine acting as a primary reinforcer (Chaudhri et al. 2007; Donny et al. 2003; Palmatier et al. 2006; Sorge et al. 2009). For example, Sorge et al. (2009) allowed rats to acquire nicotine self-administration (15 μg/kg/infusion; infusions delivered over 30 s). Though few infusions were earned across the sessions and rats acquired very slowly, acquisition criterion was met without additional cues to support behavior (Sorge et al. 2009). However, the dose range that supports self-administration is narrow, relatively few infusions are earned, and the rate of behavior does not increase in proportion to changes in the schedule of reinforcement (Caggiula et al. 2002a; Chaudhri et al. 2005, 2007; Donny et al. 2003). Certainly, the bulk of the reinforcing actions of nicotine in typical self-administration procedures cannot be explained as primary reinforcement alone. The next section will begin to dissect the role of environmental cues and their importance for maintaining smoking behavior.

4 Associative Learning and the Influence of Stimuli Predicting the Effects of Nicotine

As a consequence of the relatively weak reinforcing effects of nicotine alone, most self-administration procedures use cues paired with nicotine delivery. The use of cues is not an inherent flaw in the rodent model of human smoking, as all nicotine self-administration in humans is cued in some way. These environmental stimuli can function in multiple roles including as conditioned stimuli that trigger conditioned responses (CRs), conditioned reinforcers, and discriminative stimuli (see chapter entitled Neurobiological Bases of Cue- and Nicotine-induced Reinstatement of Nicotine Seeking: Implications for the Development of Smoking Cessation Medications; this volume).

4.1 Conditioned Responses to Nicotine-Associated Stimuli

As a result of Pavlovian conditioning, stimuli paired with nicotine can elicit CRs (Pavlov 1927). A typical Pavlovian conditioning preparation involves an existing reflexive relationship between a stimulus (unconditioned stimulus, US) and response (unconditioned response, UR). In this case, nicotine (US) results in a wide range of behavioral and physiological changes (UR). Then, an originally neutral stimulus is paired with the presentation of nicotine. Following one or more pairings, the originally neutral stimulus elicits a CR in the absence of the US and is now called a CS. In this case, any environmental stimulus that is consistently predictive of nicotine may come to serve as a CS and elicit a CR. In an operant procedure, the CS might be a light or a tone paired with nicotine delivery (provided the cue does not initially by itself elicit the response). In the context of cigarette smoking, the look or feel of a cigarette, the taste of tobacco, lighters, the cigarette pack, the smoking corner outside of the office, the effects of alcohol, certain friend groups, etc., might all function to elicit CRs, even in the absence of subsequent nicotine delivery, because they have previously been repeatedly paired with nicotine.

Evidence that stimuli associated with nicotine can elicit a CR come from a wealth of both animal and human literatures suggesting that smoking stimuli increase craving or desire to smoke cigarettes (Conklin and Tiffany 2002b; Lazev et al. 1999; Wertz and Sayette 2001). This craving is often considered to be a CR that results from the pairing of these stimuli with nicotine delivery (Conklin and Tiffany 2002a). Evidence from our laboratory has shown that tolerance to the antinociceptive and corticosterone (CORT) elevating effects of nicotine can be abolished if nicotine is delivered in a novel context (Caggiula et al. 2002b). These data suggest that the tolerance that develops is a result of a CR to the environmental stimuli that reliably predict nicotine delivery. When these environmental stimuli are absent, there is no CR and nicotine has the same effect as the first administration. This study highlights an important issue: While a CR is often similar to the UR, it can sometimes be in the opposite direction, called a compensatory CR (Siegel 1988). In this example, nicotine (US) results in antinociceptive and CORT elevation (UR). The stimuli (CS) that are paired with nicotine (US) may result in a CR opposite of that effect, so that nicotine delivery produces a smaller change than it would have acutely.

Nicotine-associated stimuli result in CRs that increase the likelihood of engaging in smoking behavior both while nicotine is being actively self-administered and as a trigger to reinitiate nicotine seeking. In relation to the former, it is well accepted that the presentation of cues paired with reward results in increased responding for the reward, even if they were never presented contingent upon the response (Rescorla and Solomon 1967), a phenomenon known as Pavlovian to instrumental transfer (see chapter entitled A Hierarchical Instrumental Decision Theory of Nicotine Dependence; volume 23). This phenomenon has received relatively little attention in animal models of nicotine self-administration, but would suggest that the mere presence of CS might facilitate both nicotine taking and other forms of reinforced behavior. The more common conceptualization of how CS influence nicotine seeking occurs during abstinence or following extinction when the CS trigger a motivational state or action schema that can lead to the experience of craving (Berridge and Robinson 2003; Tiffany, 1990). Indeed, cue-elicited craving is reliably linked to smoking in abstinent individuals (Sayette and Tiffany 2013), although this effect is not clear when assessed in non-abstinent smokers (Perkins 2009). Hence, CS can elicit a wide range of CRs including responses that may impact the probability or intensity of nicotine-seeking behavior.

4.2 Nicotine-Associated Stimuli as Conditioned Reinforcers

Because smoking stimuli have been paired with nicotine, and nicotine functions as an unconditioned (i.e., primary) reinforcer, these cues can come to reinforce behavior on their own (i.e., become conditioned reinforcers). In a stringent test of this effect, rats with a history self-administering nicotine paired with a CS learned to perform a novel response that was only reinforced by the CS (Palmatier et al. 2007). Likewise, rodent self-administration research has demonstrated that the continued delivery of cues after nicotine has been removed will maintain responding, and this rate of behavior is higher than if cues are also removed (Markou and Paterson 2009). Clinical research confirms that, over the course of a week or so, smokers will continue to smoke low-nicotine-content cigarettes (Donny et al. 2007; Donny and Jones 2009), with moderate to no decrease in smoking behavior. Furthermore, denicotinized cigarettes have been shown to substitute for nicotine-containing cigarettes better than nicotine gum (Johnson et al. 2004), and the delivery of smoking stimuli has been shown to increase ratings of liking and satisfaction better than (iv) nicotine (Rose et al. 2000). These data clearly indicate that these cues associated with nicotine can reinforce behavior.

Likewise, clinical experimental studies show that individuals who try to refrain from smoking are significantly more likely to lapse regardless of the nicotine content of those cigarette; even cigarettes with very little nicotine increased the probability of relapse compared to not smoking (Juliano et al. 2006). These data indicate that interacting with smoking stimuli may precipitate smoking behavior during abstinence. The presentation of nicotine-associated cues has also been shown to increase previously extinguished self-administration behavior in experimental animals in a phenomenon known as cue-induced reinstatement (LeSage et al. 2004). Cue-induced reinstatement is very robust; indeed, the magnitude of reinstatement is greater when it is induced by cue presentation than by nicotine (LeSage et al. 2004). Upon cessation of nicotine use, cues should eventually undergo extinction as they no longer reliably predict nicotine delivery (Caggiula et al. 2001; Cohen et al. 2005; Liu et al. 2007, 2008, 2010). However, extinction is context dependent (Wing and Shoaib 2008; Bouton 2011), so extinction learning would need to occur in multiple contexts before cues would be fully extinguished, a potentially lengthy process.

4.3 Nicotine-Associated Contexts as Discriminative Stimuli

Smoking cues and contexts can also serve as discriminative stimuli, which tend to be broader contextual stimuli, signaling when behavior will result in reinforcement. In a typical discrimination preparation, individuals learn that a behavior will result in reinforcement in the presence of one stimulus and will not result in reinforcement in the absence of that stimulus (Skinner 1953). For learning to take place, the probability of reinforcement in the presence of the stimulus must be greater than the probability of reinforcement when the stimulus is absent. In self-administration paradigms, rodents undoubtedly learn that the operant chamber signals the availability of nicotine, although rodents unavoidably spend some time in the chamber when nicotine is not available (right before the session is started, right after the session ends, during time out periods post-infusion when the drug may no longer be available). Some researchers use other stimuli (cue lights, house lights, tones) to signal the start of the session or to signal "time in" (Belluzzi et al. 2005;

Grebenstein et al. 2013; Hall et al. 2014). The presence of these stimuli will increase the likelihood of engaging in the reinforced response. This issue is considered in more detail elsewhere in the book.

These contextual stimuli make conducting human laboratory smoking research difficult because smoking behavior has not been previously reinforced in the laboratory context and conclusions drawn from studies conducted in laboratory environments may not extend to the natural environment. In two studies, smokers were asked to smoke cigarettes with very low nicotine contents. In a study conducted in an in-patient hospital unit, there was a decrease in smoking behavior when smokers switched to these low-nicotine cigarettes (Donny et al. 2007). However, smoking behavior did not change when a similar study was conducted in the natural environment (Donny and Jones 2009). These data parallel the work by Wing and Shoaib (2008) in which extinction was shown to proceed more quickly in a novel environment not associated with previous nicotine self-administration. It may have been easier for smokers to learn that smoking no longer resulted in nicotine delivery in a completely novel context than the natural environment in which smoking behavior had a long history of being reinforced across many contexts (Conklin 2010; Wray et al. 2011).

4.4 Individual Differences in Associative Learning Effects

There is likely a large degree of variability in the degree to which these cues are involved in smoking behavior between individuals. One theory posits that all individuals learn about cues in their environment, but there is variability in the degree to which these cues are "wanted" (Robinson and Berridge 2000; Robinson et al. 2014). In rodent models, animals that show attraction toward cues have higher break points for cocaine on progressive ratio schedules (generally considered to be an indicator of motivation to obtain drug) (Saunders and Robinson 2011), the development of cocaine-paired CPP (Meyer et al. 2011), greater cocaine sensitization upon repeat treatment (Flagel et al. 2008), and greater cue-induced reinstatement (Yager and Robinson 2010). Future research may provide important insight into the variability observed in nicotine self-administration and the role of cues following nicotine reduction. Sex may be another determinant of cue effects. On average, women are more likely to be affected by smoking cues than men, who may be more directly influenced by nicotine (Perkins 2009; Perkins et al. 2002). However, preclinical work does not support this idea. One study evaluated cue-induced and nicotine-primed reinstatement in male and female rats (Feltenstein et al. 2012). There were no sex- or estrous cycle-dependent differences between male and female rats. Still, expanding this line of research is important for cessation, where women may have less success with nicotine replacement therapy because smoking stimuli are more critical in the maintenance of smoking behavior (Perkins et al. 2002).

5 Associative Learning and the Influence of Nicotine as a Predictor of Other Reinforcers

A very different line of research investigates the role of nicotine as a cue for other stimuli. For example, rats can learn to respond on one lever for food if they have received an injection of nicotine and respond on another lever if they received a saline injection (Stolerman 1989). In these experiments, nicotine signals that one behavior will be reinforced and another will not. Relatedly, humans can learn to engage in one response if they receive nicotine nasal spray and another response if they receive saline spray (Perkins et al. 1994). Researchers have extended this finding to show that nicotine, when paired with a reinforcer, even when it is not contingent upon behavior, can increase the rate of behavior directed at the location of reinforcer delivery (i.e., goal tracking; Besheer et al. 2004).

The role of nicotine as a predictor of other reinforcers in acquiring or maintaining smoking behavior is unclear. However, nicotine delivery is paired with many reinforcing stimuli in the natural environment, and nicotine may function as a CS and a conditioned reinforcer, through this pairing. For example, an adolescent who receives peer acceptance when engaging in smoking behavior may experience nicotine as a conditioned reinforcer because of this repeated pairing over time (i.e., nicotine predicts peer acceptance). It is difficult to show experimentally that nicotine can acquire additional reinforcing value through this association because nicotine has existing primary reinforcing value. However, research has shown that pairing diazepam, an anxiolytic, with money can result in a preference for diazepam, highlighting the ability of drugs to acquire reinforcing value through pairing with other reinforcers (Alessi et al. 2002). Nicotine may also come to elicit CRs as a result of pairing with other drugs of abuse, although there is no existing research in this area. Drugs such as alcohol, marijuana, and caffeine are frequently co-used with nicotine. Smokers who have a cigarette with their morning coffee may associate nicotine with coffee and enjoy cigarettes even on mornings when they choose decaffeinated coffee.

6 Reinforcement-Enhancing Effects of Nicotine

The primary reinforcing effects of nicotine and the consequent associative and conditioned reinforcing properties of nicotine are important in driving smoking behavior, but additional non-associative effects of nicotine on reinforced behavior may be equally important. Nicotine potentiates the reinforcing properties of other rewards. This latter effect occurs independent of any predictive relationship between nicotine and either the other stimulus or the target behavior (Caggiula et al. 2009).

In the first study to describe the effect of nicotine on responding maintained by other stimuli in the environment, male adult rats were allowed to respond for nicotine, saline, and/or a VS (VS; the 1-s onset of a cue light above the active lever and the 60-s offset of the chamber light) in a between groups design (Donny et al. 2003). Response-contingent nicotine, by itself, results in very low rates of responding and, in some cases, failed to support self-administration. Interestingly, responding for the VS, which has been used previously as a drug cue (Corrigall and Coen 1989; Donny et al. 1995), supported behavior even in the absence of nicotine, suggesting the VS functioned as an unconditioned reinforcer. This finding is consistent with data describing that sensory stimuli can function as reinforcers (Fowler 1971, Harrington 1963). Importantly, pairing nicotine with the VS produced a synergistic, not just additive, effect on behavior. Responding was more than twice the sum of response rates produced either by nicotine alone or the VS alone. This is consistent with previous studies demonstrating the importance of environmental cues in nicotine self-administration (Caggiula et al. 2001, 2002a, b; Cohen et al. 2005), but also raised questions about the nature of the relationship between nicotine and the VS. The synergistic effects seemed disproportional with the reinforcing properties of the two stimuli. However, this study included a critical control condition that provided a potential answer to these questions. In this group, animals were allowed to respond for the VS while receiving infusions of nicotine that were controlled by (yoked to) another animal. Remarkably, there were no differences in responding for the VS during acquisition between the contingent and non-contingent nicotine conditions. Both contingent and non-contingent infusions of nicotine resulted in the synergistic enhancement of responding for a VS compared to responding for the VS with saline infusion (Donny et al. 2003). Therefore, the increase in responding by the pairing of nicotine with the VS could not be explained by the VS functioning as a CS.

Enhancement of the reinforcement by nicotine is entirely consistent with extensive data on the effects of nicotine, and other drugs of abuse, on intracranial self-stimulation (ICSS) and CPP. In ICSS studies, a stimulating electrode is typically placed in the posterior lateral hypothalamus and rats respond for brain stimulation reward. Systemic (Harrison 2002) and self-administered (Kenny and Markou 2006) nicotine lower the threshold for ICSS, indicating that nicotine enhances the rewarding properties of brain stimulation. Likewise, early studies demonstrated that psychostimulants enhance responding for conditioned stimuli (Beninger 1980; Robbins and Koob 1978). Although these studies were never extended to nicotine, they highlight that these effects are not unique to nicotine, an observation we have also made (Chaudhri et al. 2006b).

6.1 *Reinforcement-Enhancing Effects of Nicotine Occur Across Routes of Administration*

It is possible that the pattern of nicotine delivery may affect responding for the VS. In particular, yoked infusions of nicotine might result in intermittent unintentional pairings between nicotine and responding for the VS, leading to an associative relationship. To test whether the observed enhancement was due to spurious associations, responding for the VS was tested in rats receiving either rapid non-contingent infusions independent of responding for the VS, a constant nicotine infusion over the 1-h period, or contingent nicotine infusion (Donny et al. 2003). Despite the varied patterns of nicotine delivery, all rats had similar elevated levels of responding for VS presentations compared to saline. Similarly, systemic injection of nicotine administered before the session and even acute treatment with osmotic minipumps enhances responding for VS presentations. These data support the hypothesis that nicotine can directly enhance behavior maintained by unconditioned non-pharmacological reinforcers and that the enhancement is not dependent on a discrete temporal relationship with the stimulus or the behavior. Importantly, replacing nicotine with saline resulted in an immediate reduction in responding to levels similar to controls. Furthermore, reinstating nicotine to contingent and non-contingent groups immediately increased responding to pre-extinction levels (Chaudhri et al. 2007; Donny et al. 2003; Palmatier et al. 2007). These basic observations have now been replicated many times across a range of doses, routes of administration, schedules of reinforcement, and reinforcing stimuli, including conditioned reinforcers (Chaudhri et al. 2005, 2006a; Donny et al. 1999; Palmatier et al. 2006).

To further examine the relative contribution of the primary reinforcing and the reinforcement enhancement properties of nicotine, we utilized a paradigm in which rats had the option to respond for nicotine and the VS independently (Palmatier et al. 2006). The primary group of interest pressed one active lever for the presentation of the VS and another lever for infusions of nicotine. A separate group of rats (nicotine + VS) had standard self-administration operanda: One active lever controlled both the presentation of the VS and nicotine infusion. Control groups received either nicotine infusions (nicotine only) or VS presentations (VS only) contingent upon behavior on the active lever (the other lever was inactive). Nicotine alone and VS alone maintained relatively moderate levels of behavior and access to both reinforcers on one lever resulted in synergistic enhancement of responding, as we had previously reported. Surprisingly, when rats had access to each reinforcer on a separate lever, responding on the nicotine lever was low, about equivalent level to that of the nicotine only group. However, responding for the VS lever was enhanced by the same magnitude as the nicotine + VS group. Additional studies from our laboratory have demonstrated that the metabotropic glutamate five receptor antagonist MTEP can suppress responding for nicotine alone, with no effect on the reinforcement-enhancing properties of nicotine, indicating that the primary and enhancing reinforcement properties of nicotine can be pharmacologically dissociated (Palmatier et al. 2008). To our knowledge, there are no data indicating what neural mechanism(s) may be responsible for the reinforcing enhancement properties of nicotine. These collective studies indicate that the reinforcement-enhancing properties of nicotine are potent even when only a small amount of the drug is administered that the effects are behaviorally and pharmacologically dissociable and that the high rates of self-administration observed in the paired group is likely due to the reinforcement-enhancing effects of nicotine (Palmatier et al. 2006, 2008).

Additional studies by Chaudhri et al. (2006a) demonstrated that in rats that acquired self-administration across a range of nicotine doses in the absence of a pairing with a non-pharmacological stimulus, the addition of the VS resulted in an immediate and robust increase in responding. These results are important because they (1) replicate the observation that rats will acquire responding for nicotine self-administration without coincident non-pharmacological stimuli, but that this effect depends on larger doses of nicotine, and (2) demonstrate that the effect of the addition of the VS was most prominent at low doses of nicotine.

Taken together, these studies emphasize why responding for nicotine infusions is so robust in the presence of a non-drug stimulus: It is not the nicotine per se but the synergistic interaction between nicotine and non-pharmacological stimuli that produces a significant increase in behavior. Environmental stimuli with reinforcing value paired with nicotine increase the rate of acquisition of nicotine self-administration and the rate of maintained self-administered behavior independent of the route and speed of nicotine delivery (Caggiula et al. 2009; Chaudhri et al. 2006a).

6.2 Reinforcement-Enhancing Effects of Nicotine Occur Across Age and Sex

Until recently, the majority of research on the reinforcement-enhancing effects of nicotine has been conducted in adult male rats. However, the enhancing effects have been demonstrated in adolescent male rats as well. Work from our laboratory assessed whether responding for the VS was enhanced by exposure to nicotine in adolescent (postnatal day 29–42), male rats (Weaver et al. 2012). Like adults, adolescent rats responded for presentations of the VS and subcutaneous (sc) nicotine just prior to the session increased responding for the VS. The effect was qualitatively similar to that observed in adults at the dose tested. Similarly, adolescent male rats (P39–40) tested in a CPP procedure using social reward (i.e., a "playmate") found that nicotine increased the amount of time spent on the side of the chamber paired with a social playmate (Theil et al. 2009). To date, potential developmental differences in the reinforcement-enhancing effects of nicotine have not been thoroughly examined.

Another important question given that female rats acquire nicotine self-administration more quickly and reach higher break points on a PR schedule when nicotine is paired with the VS (Donny et al. 2000; Lanza et al. 2004) is whether there are sex differences in the reinforcement-enhancing effects of nicotine. To our knowledge, this question has not been directly addressed. Studies suggest, however, that there are sex differences in self-administration of both nicotine alone and in combination with the VS. In a study by Chaudhri et al. (2005), animals first acquired responding for contingent nicotine infusions without additional non-pharmacological stimuli. Females earned more infusions of moderate to large (60–120 μg/kg) doses of nicotine than males. Then, the VS was added with active lever presses resulting in delivery of both the nicotine infusion and the VS. The addition of VS caused a doubling in responding in both sexes when animals were self-administering 30 or 60 μg/kg/infusion (Chaudhri et al. 2005). A separate group of male and female rats were allowed to respond for VS presentations without a nicotine infusion and confirmed that the VS functioned as a primary reinforcer in females as well as males, suggesting the potentiation in self-administration may have been driven by the reinforcement-enhancing effects of nicotine in both sexes.

6.3 Reinforcement-Enhancing Effects of Nicotine Occur in Humans

The ability of nicotine to enhance reinforcement for other stimuli is well established in the rodent model, but whether reinforcement enhancement is important in human smokers has received less attention. Studies to date, however, are consistent with the animal literature. Smoking increased ratings of attractiveness (Attwood et al. 2009) and reported feelings of happiness during happy films (Dawkins et al. 2007). Likewise, transdermal nicotine increases the response bias toward a rewarded stimulus (Barr et al. 2008). Furthermore, abstinence from smoking reduces the blood-oxygen-level-dependent response to monetary rewards in the caudate. Finally, in the most direct study of reinforcement-enhancing effects in humans, Perkins and Karelitz evaluated the ability of nicotine to enhance operant responding for a variety of rewards in smokers (Perkins and Karelitz 2013a). Participants were dependent and non-dependent smokers that were deprived of nicotine at the start of the study and asked to respond on a PR schedule adapted for human subjects for a designated music reward. Smoking was able to enhance responding for high preference music, as compared to no smoking and smoking at low levels that are subthreshold for enhancement (Perkins and Karelitz 2013a). The implications of these results are discussed in detail below, but they provide clear evidence that nicotine enhances reinforcement in humans.

6.4 Enhancement is Moderated by the Type and Nature of the Reinforcer

If a stimulus has little or no reinforcing value (i.e., neutral), nicotine should have no effect on enhancing the rewarding properties of that stimulus unless it gains conditioned reinforcing value (as discussed above). In a study designed to test this hypothesis, we compared the effects of systemic non-contingent injections on two sensory stimuli that differed in their unconditioned reinforcing effects. During sessions of acquired stable behavior, rats responded significantly more for a houselight-off stimulus (5-s extinction of the houselight paired with an 83-dB tone) than a lever light-on stimulus (5-s onset of a stimulus light above the active lever paired with an 83-dB tone), indicating that the houselight-off stimulus has higher incentive value.

Saline injections had no effect on behavior. Nicotine caused an increase in responding for the houselight-off stimulus, but with no effect on the lever light-off stimulus. These results support the notion that nicotine has the ability to non-associatively enhance responding only for stimuli with reinforcing value.

More recent studies have further confirmed our hypothesis that the magnitude of the nicotine-induced enhancement is modulated by the strength of the reinforcer. The quantitative relationship between the enhancing effects of nicotine and the incentive value of the reward was tested using rats responding on a PR schedule of reinforcement for the delivery of liquid sucrose (0, 5, 20, and 60 %; Palmatier et al. 2012). Systemic, non-contingent injection of nicotine enhanced the responding for sucrose reward. The break point reached (final number of responses required for reward delivery) was potentiated by nicotine, with the magnitude of enhancement increasing with increasing sucrose concentration. These results are particularly interesting, as they raise the possibility that the reward enhancement effects of nicotine may override other pharmacological actions of nicotine, as nicotine is known to be a potent appetite suppressant. In fact, when the sucrose solutions (2.5, 5, and 10 %) were available on a reinforcement schedule that demanded low responding rates for each reward delivery (FR 3), nicotine reduced responding. As has been previously suggested, the effects of nicotine may dependent on the schedule of reinforcement because nicotine has multiple actions on food reinforcement; nicotine may enhance satiety when food is relatively freely available and enhance the reinforcing efficacy of food when it is relatively restricted (Donny et al. 2011).

Work from our laboratory has also established that nicotine has the ability to enhance responding for conditioned reinforcers. Work by Chaudhri et al. 2006a, b used a light-tone stimulus as a CS predictive of sucrose pellet delivery. Both contingent and non-contingent delivery of nicotine enhanced responding for a light-tone stimulus when it had been previously paired with sucrose, but not if the stimulus was not predictive of sucrose. Extending this work to a nicotine-associated CS, Palmatier et al. (2007) trained animals to lever press for nicotine paired with as unconditioned light stimulus. Animals were then allowed to nosepoke for the CS alone and demonstrated the predicted conditioned reinforcing effects (as described above). Interestingly, non-contingent delivery of nicotine further increased the rate of responding for the CS (Palmatier et al. 2007), confirming that non-contingent nicotine can enhance responding that is maintained solely by a nicotine-associated conditioned reinforcer.

Work investigating the ability of nicotine to enhance reward in humans has found similar results. One study tested whether smoking enhanced responding for music reward on a PR schedule (Perkins and Karelitz 2013a). Participants were instructed to bring a pack of their own cigarettes and a music album of their choice. Different music was designated as high, moderate, or low reward based on the participants own rating of the music on a 0–100 visual analog scale (VAS). Participants completed three 2-h sessions, each after overnight abstinence, with different levels of smoking prior to assessing responding to hear different segments of music. As our hypothesis would predict, smoking enhanced responding and this was only observed when music was preferred (Perkins and Karelitz 2013a).

Minimal (couple puffs) or no smoking had no effect on behavior. Hence, the effect of nicotine on reinforced behavior in humans depends both on the level of smoke exposure and the degree to which the reward is preferred.

Furthermore, in a separate study investigating that ability of nicotine to enhance responding for reward in humans, smoking after abstinence was able to enhance responding for a preferred music reward, but not for monetary reward (Perkins and Karelitz 2013b). These data suggest that the reinforcement-enhancing effects of nicotine differ depending on reinforcer type (e.g., more apparent with "sensory" reinforcers). However, it is possible that the small monetary rewards (participants earned an average of about $1.20 in a 2-h session) were effectively neutral and thus not able to be enhanced by nicotine.

Together, these studies provide evidence that the reinforcement-enhancing effects of nicotine are dependent upon the reinforcer being presented (Barret and Bevins 2013; Palmatier 2012; Perkins and Karelitz 2013b). Neutral or mild reinforcers as less likely to be impacted by nicotine unless the value of those reinforcers is increased as a consequence of being paired with another reinforcer. In addition, some reinforcers may be more prone to the reinforcement-enhancing effects of nicotine or be masked by other effects of nicotine in some conditions, for example, "sensory" reinforcers and food, respectively.

6.5 Implications of the Reinforcement-Enhancing Effect of Nicotine for Tobacco and Other Nicotine-containing Products

In comparison with the primary reinforcing actions of nicotine, the reinforcement enhancement actions of nicotine are seen with lower doses and drug delivery that is not temporally tied to behavior. Thus, for example, patch delivery of nicotine might enhance other reinforcers (and may contribute to its efficacy as a smoking cessation aid) but is unlikely to have primary reinforcing actions (Barr et al. 2008). Similarly, new tobacco products that provide nicotine delivery that is neither rapid nor discrete will likely favor the reinforcement enhancement actions of nicotine as opposed to the primary reinforcing actions. E-cigarettes, for example, are perceived favorably by young adults, especially when appealing flavorants are added to the product (Choi et al. 2012). Given the ability of nicotine to enhance responding for sucrose reward (Barret and Bevins 2013; Palmatier et al. 2012; Schassburger et al. 2013), it is possible that nicotine by "vaping" an e-cigarette might enhance the reinforcing properties of the flavorant within that e-cigarette.

Conversely, understanding the potentially different neuropharmacological mechanisms underlying the primary reinforcing and the reinforcement-enhancing actions of nicotine may impact the development of smoking cessation pharmacotherapies. Interestingly, both bupropion (Caggiula et al. 2009) and varenicline (Levin et al. 2012), both FDA-approved prescription pharmacotherapies for the treatment of smoking cessation, can substitute for the reinforcement-enhancing effects of nicotine, which may partially underlie their efficacy as pharmacotherapies. To test the ability of bupropion to enhance the reinforcing valence of VS presentations, nicotine or bupropion was systemically administered across sessions (Palmatier et al. 2009). Both nicotine and bupropion increased responding for VS presentations. As expected, the nicotine enhancement was abolished by the administration of mecamylamine, a non-selective nicotinic antagonist. Mecamylamine had no effect on the enhancement caused by buproprion; the bupropion-induced enhancement was blocked by the administration of an alpha adrenergic receptor antagonist, indicating that the reinforcement-enhancing effects of these drugs are pharmacologically dissociable. Levin et al. (2012) tested the effects of systemic administration of varenicline, a partial nicotinic agonist, on responding for VS presentations with and without co-administration with nicotine (Levin et al. 2012). Varenicline dose dependently increased VS presentations earned, as well as suppressed nicotine-induced enhancement. As current over-the-counter nicotine replacement therapies are largely ineffective tools in supporting smoking cessation (Kotz et al. 2014), it might be more beneficial to target the development of new medications at the reinforcement-enhancing actions of nicotine, as that seems to be a primary mechanism for nicotine self-administration.

7 Summary and Conclusions

Nicotine reinforcement is remarkably complex and requires understanding of both associative and non-associative mechanisms well beyond the primary reinforcing effects of the drug itself. From an associative perspective, environmental stimuli that predict nicotine delivery can become conditioned stimuli eliciting CRs that increase the probability of smoking. They will also become conditioned reinforcers through their pairing with nicotine, an unconditioned reinforcer, and will reinforce smoking behavior on their own. Contextual stimuli can become discriminative stimuli, increasing the probability of engaging in smoking behavior. Nicotine can also enter into associative relationships through its pairing with other reinforcing stimuli in the environment and consequently function as a CS or as a conditioned reinforcer, both of which may increase the likelihood of engaging in smoking behavior.

From a non-associative perspective, nicotine, like other psychostimulants, can directly impact reinforcement from other stimuli in the environment. This effect is particularly robust with nicotine and has been emphasized in the "dual-reinforcement" model, which posits that nicotine maintains self-administration behavior as a primary reinforcer and a reinforcement enhancer (Caggiula et al. 2009; Donny et al. 2003). The reinforcement-enhancing properties of nicotine have been observed in experimental animals across age and sex, and, more recently, confirmed to impact behavior in humans. Responding for both unconditioned reinforcers (sensory stimuli, food reward) and conditioned reinforcers (nicotine and non-nicotine related) can be enhanced by nicotine. The magnitude of the enhancement is dependent on the magnitude of the reinforcer and potentially the type of reinforcer and conditions under which it is available.

In sum, self-administration of nicotine in humans and the rodent model is sustained by three main actions: (1) nicotine acts as a relatively weak primary reinforcer; (2) nicotine can establish conditioned reinforcers in the environment through associative processes; and (3) nicotine can potentiate the incentive valence of other stimuli with reinforcing value. Other actions may also be important (e.g., nicotine as a CS), but future studies will be needed to confirm these effects. These studies confirm that nicotine is the key psychoactive determinant of tobacco product use; however, it is much more insidious than might be expected. Thus, understanding how nicotine acts to maintain behavior is still at the heart of reducing the burden of tobacco dependence.

Acknowledgments Research reported in this publication was supported by the National Institute on Drug Abuse (R01 DA010464) and FDA Center for Tobacco Products (CTP) (U54 DA031659). The funding source had no other role other than financial support. The content is solely the responsibility of the authors and does not necessarily represent the official views of the NIH or the Food and Drug Administration.

References

Abelson PH (1995) Flaws in risk assessments. Science 270:215

Adriani W, Spijker S, Deroche-Gamonet V, Laviola G, Le Moal M, Smit AB, Piazza PV (2003) Evidence for enhanced neurobehavioral vulnerability to nicotine during periadolescence in rats. J Neurosci 23:4712–4716

Adriani W, Deroche-Gamonet V, Le Moal M, Laviola G, Piazza PVV (2006) Preexposure during or following adolescence differently affects nicotine-rewarding properties in adult rats. Psychopharmacology 184:382–390

Amit Z, Brown Z, Rockman G (1977) Possible involvement of acetaldehyde, norepinephrine and their tetrahydroisoquinoline derivatives in the regulation of ethanol self-administration. Drug Alcohol Depend 2:495–500

Ator NA, Griffiths RR (2003) Principles of drug abuse liability assessment in laboratory animals. Drug Alcohol Depend 70:S55–S72

Attwood AS, Penton-Voak IS, Munafo MR (2009) Effects of acute nicotine administration on ratings of attractiveness of facial cues. Nicotine Tob Res 11:44–48

Bardo MT, Green TA, Crooks PA, Dwoskin LP (1999) Nornicotine is self-administered intravenously by rats. Psychopharmacology 146:290–296

Barr RS, Pizzagalli DA, Culhane MA, Goff DC, Evins AE (2008) A single dose of nicotine enhances reward responsiveness in nonsmokers: implications for development of dependence. Biol Psychiatry 63:1061–1065

Barret ST, Bevins RA (2013) Nicotine enhances operant responding for qualitatively distinct reinforcers under maintenance and extinction conditions. Pharmacol Biochem Behav 114–115:9–15

Becker JB, Hu M (2008) Sex differences in drug abuse. Front Neuroendocrinol 29:36–47

Belluzzi JD, Lee AG, Oliff HS, Leslie FM (2004) Age-dependent effects of nicotine on locomotor activity and conditioned place preference in rats. Psychopharmacology 174:389–395

Belluzzi JD, Wang R, Leslie FM (2005) Acetaldehyde enhances acquisition of nicotine self-administration in adolescent rats. Neuropsychopharmacology 30:705–712

Beninger RJ, Hanson DR, Phillips AG (1980) The acquisition of responding with conditioned reinforcement: effects of cocaine, (+)-amphetamine and pipradrol. Br J Pharmacol 1:149–154

Benowitz NL (1990) Pharmacokinetic considerations in understanding nicotine dependence. In: The biology of nicotine dependence, Ciba Foundation Symposium, Wiley, New York

Benowitz NL, Hatsukami D (1998) Gender differences in the pharmacology of nicotine addiction. Addict Biol 3:383–404

Berchtold NC, Cribbs DH, Coleman PD, Rogers J, Head E, Kim R, Beach T, Miller C, Troncoso J, Trojanowski JQ (2008) Gene expression changes in the course of normal brain aging are sexually dimorphic. Proc Nat Acad Sci 105:15605–15610

Berlin I, Said S, Spreux-Varoquaux O, Olivares R, Launay JM, Puech AJ (1995) Monoamine oxidase A and B activities in heavy smokers. Biol Psychiatry 38:756–761

Berridge KC, Robinson TE (2003) Parsing reward. Trends Neurosci 26:507–513

Besheer J, Palmatier MI, Metschke DM, Bevins, RA (2004) Nicotine as a signal for the presence or absence of sucrose reward: a Pavlovian drug appetitive conditioning preparation in rats. Psychopharmacology 172:108–117

Bevins RA (2009) Altering the motivational function of nicotine through conditioning processes. Nebr Symp Motiv 55:111–129

Bouton ME (2011) Learning and the persistence of appetite: extinction and the motivation to eat and overeat. Physiol Behav 103:51–58

Brady KT, Randall CL (1999) Gender differences in substance use disorders. Psychiatr Clin North Am 22:241–252

Breslau N, Peterson EL (1996) Smoking cessation in young adults: age at initiation of cigarette smoking and other suspected influences. Am J Public Health 86:214–220

Brielmaier JM, McDonald CG, Smith RF (2007) Immediate and long-term behavioral effects of a single nicotine injection in adolescent and adult rats. Neurotoxicol Teratol 29:74–80

Brown Z, Amit Z, Rockman G (1979) Intraventricular self-administration of acetaldehyde, but not ethanol, in naive laboratory rats. Psychopharmacology 64:271–276

Busto U, Sellers EM (1986) Pharmacokinetic determinants of drug abuse and dependence. A conceptual perspective. Clin Pharmacokinet 11:144–153

Caggiula AR, Donny EC, White AR, Chaudhri N, Booth S, Gharib MA, Sved AF (2001) Cue dependency of nicotine self-administration and smoking. Pharmacol Biochem Behav 70:515–530

Caggiula AR, Donny EC, Chaudhri N, Perkins KA, Evans-Martin FF, Sved AF (2002a) Importance of nonpharmacological factors in nicotine self-administration. Physiol Behav 77:683–687

Caggiula AR, Donny EC, White AR, Chaudhri N, Booth S, Gharib MA, Sved AF (2002b) Environmental stimuli promote the acquisition of nicotine self-administration in rats. Psychopharmacology 163:230–237

Caggiula AR, Donny EC, Palmatier MI, Liu X, Chaudhri N, Sved AF (2009) The role of nicotine in smoking: a dual-reinforcement model. Nebr Symp Motiv 55:91–109

Caine SB, Collins GT, Thomsen M, Wright C, Lanier RK, Mello NK (2014) Nicotine-like behavioral effects of the minor tobacco alkaloids nornicotine, anabasine, and anatabine in male rodents. Exp Clin Psychopharmacol 22:9–22

Carroll ME, Lynch WJ, Roth ME, Morgan AD, Cosgrove KP (2004) Sex and estrogen influence drug abuse. Trends Pharmacol Sci 25:273–279

CDC (2010) Cigarette use among high school students-United States, 1991–2009. Morb Mortal Wkly Rep (MMWR) 59:797

CDC (2012) Current cigarette smoking among adults-United States, 2011. Morb Mortal Wkly Rep (MMWR) 61:889–894

Chambers RA, Taylor JR, Potenza MN (2003) Developmental neurocircuitry of motivation in adolescence: a critical period of addiction vulnerability. Am J Psychiatry 160:1041–1052

Chaudhri N, Caggiula AR, Donny EC, Booth S, Gharib MA, Craven LA, Perkins KA (2005) Sex differences in the contribution of nicotine and nonpharmacological stimuli to nicotine self-administration in rats. Psychopharmacology 180:258–266

Chaudhri N, Caggiula AR, Donny EC, Booth S, Gharib M, Craven L, Sved AF (2006a) Operant responding for conditioned and unconditioned reinforcers in rats is differentially enhanced by the primary reinforcing and reinforcement-enhancing effects of nicotine. Psychopharmacology 189:27–36

Chaudhri N, Caggiula AR, Donny EC, Palmatier MI, Liu X, Sved AF (2006b) Complex interactions between nicotine and nonpharmacological stimuli reveal multiple roles for nicotine in reinforcement. Psychopharmacology 184:353–366

Chaudhri N, Caggiula AR, Donny EC, Booth S, Gharib M, Craven L, Sved SF (2007) Self-administered and noncontingent nicotine enhance reinforced operant responding in rats: impact of nicotine dose and reinforcement schedule. Psychopharmacology 190:353–362

Chen H, Matta SG, Sharp BM (2007) Acquisition of nicotine self-administration in adolescent rats given prolonged access to the drug. Neuropsychopharmacology 32:700–709

Cheskin LJ, Hess JM, Henningfield J, Gorelick DA (2005) Calorie restriction increases cigarette use in adult smokers. Psychopharmacology 179:430–436

Choi K, Fabian L, Mottey N, Corbett A, Forster J (2012) Young adults' favorable perceptions of snus, dissolvable tobacco products, and electronic cigarettes: findings from a focus group study. Am J Public Health 102:2088–2093

Clemens KJ, Caillé S, Stinus L, Cador M (2009) The addition of five minor tobacco alkaloids increases nicotine-induced hyperactivity, sensitization and intravenous self-administration in rats. Int J Neuropsychopharmacol 12:1355–1366

Cohen A, Koob GF, George O (2012) Robust escalation of nicotine intake with extended access to nicotine self-administration and intermittent periods of abstinence. Neuropsychopharmacology 37:2153–2160

Cohen C, Perrault G, Griebel G, Soubrie P (2005) Nicotine-associated cues maintain nicotine-seeking behavior in rats several weeks after nicotine withdrawal: reversal by the cannabinoid (CB1) receptor antagonist, rimonabant (SR141716). Neuropsychopharmacology 30:145–155

Conklin CA, Tiffany S (2002a) Cue-exposure treatment: time for change. Addiction 97:1219–1221

Conklin CA, Tiffany ST (2002b) Applying extinction research and theory to cue-exposure addiction treatments. Addiction 97:155–167

Conklin CA, Perkins KA, Robin N, McClernon FG, Salkeld RP (2010) Bringing real world into the laboratory: person smoking and nonsmoking environments. Drug Alcohol Depend 111:58–63

Correa M, Salamone JD, Segovia KN, Pardo M, Longoni R, Spina L, Peana AT, Vinci S, Acquas E (2012) Piecing together the puzzle of acetaldehyde as a neuroactive agent. Neurosci Biobehav Rev 36:404–430

Corrigall WA (1999) Nicotine self-administration in animals as a dependence model. Nicotine Tob Res 1:11–20

Corrigall WA, Coen KM (1989) Nicotine maintains robust self-administration in rats on a limited-access schedule. Psychopharmacology 99:473–478

Costello MR, Reynaga DD, Mojica CY, Zaveri NT, Belluzzi JD, Leslie FM (2014) Comparison of the reinforcing properties of nicotine and cigarette smoke extract in rats. Neuropsychopharmacol 39:1843–1851

Crooks PA, Li M, Dwoskin LP (1997) Metabolites of nicotine in rat brain after peripheral nicotine administration cotinine, nornicotine, and norcotinine. Drug Metab Dispos 25:47–54

Dawkins LPJ, West R, Powell J, Pickering A (2007) A double-blind placebo-controlled experimental study of nicotine: II–Effects on response inhibition and executive functioning. Psychopharmacology 190:457–467

Deng XS, Deitrich RA (2008) Putative role of brain acetaldehyde in ethanol addiction. Current drug abuse reviews 1:3–8

DeNoble VJ, Mele PC (1983) Behavioral pharmacology annual report. Philip Morris Tobacco Resolution. Bates no. 20605661. Available at http://www.pmdocs.com/getallimg.asp%3Fif%BCavpidx

Donny EC, Jones M (2009) Prolonged exposure to denicotinized cigarettes with or without transdermal nicotine. Drug Alcohol Depend 104:23–33

Donny EC, Caggiula AR, Knopf S, Brown C (1995) Nicotine self-administration in rats. Psychopharmacology 122:390–394

Donny EC, Caggiula AR, Mielke MM, Jacobs KS, Rose C, Sved AF (1998) Acquisition of nicotine self-administration in rats: the effects of dose, feeding schedule, and drug contingency. Psychopharmacology 136:83–90

Donny EC, Caggiula AR, Mielke MM, Booth S, Gharib MA, Hoffman A, McCallum SE (1999) Nicotine self-administration in rats on a progressive ratio schedule of reinforcement. Psychopharmacology 147:135–142

Donny EC, Caggiula AR, Rowell PP, Gharib MA, Maldovan V, Booth S, McCallum S (2000) Nicotine self-administration in rats: estrous cycle effects, sex differences and nicotinic receptor binding. Psychopharmacology 151:392–405

Donny EC, Chaudhri N, Caggiula AR, Evans-Martin FF, Booth S, Gharib MA, Sved AF (2003) Operant responding for a visual reinforcer in rats is enhanced by noncontingent nicotine: implications for nicotine self-administration and reinforcement. Psychopharmacology 169:68–76

Donny EC, Houtsmuller E, Stitzer ML (2007) Smoking in the absence of nicotine: behavioral, subjective and physiological effects over 11 days. Addiction 102:324–334

Donny EC, Caggiula AR, Weaver MT, Levin ME, Sved AF (2011) The reinforcement-enhancing effects of nicotine: implications for the relationship between smoking, eating and weight. Physiol Behav 104:143–148

Duffy PH, Feuers R, Hart RW (1990) Effect of chronic caloric restriction on the circadian regulation of physiological and behavioral variables in old male B6C3F1 mice. Chronobiol Int 7:291–303

Farre M, Cami J (1991) Pharmacokinetic considerations in abuse liability evaluation. Br J Addict 86:1601–1606

Feltenstein MW, Ghee SM, See RE (2012) Nicotine self-administration and reinstatement of nicotine-seeking in male and female rats. Drug Alcohol Depend 121:240–246

Flagel SB, Watson SJ, Akil H, Robinson TE (2008) Individual differences in the attribution of incentive salience to a reward-related cue: influence on cocaine sensitization. Behav Brain Res 186:48–56

Fowler CD, Kenny PJ (2011) Intravenous nicotine self-administration and cue-induced reinstatement in mice: effects of nicotine dose, rate of drug infusion and prior instrumental training. Neuropharmacology 61:687–698

Fowler HARRY (1971) Implications of sensory reinforcement. The nature of reinforcement, Academic Press, New York, 151–195

Fowler JS, Volkow ND, Wang G-J, Pappas N, Logan J, Shea C, Alexoff D, MacGregor RR, Schlyer DJ, Zezulkova I (1996a) Brain monoamine oxidase a inhibition in cigarette smokers. Proc Natl Acad Sci USA 93:14065–14069

Fowler J, Volkow N, Wang G-J, Pappas N, Logan J, MacGregor R, Alexoff D, Shea C, Schlyer D, Wolf A (1996b) Inhibition of monoamine oxidase B in the brains of smokers. Nature 379:733–736

Goldberg SR, Spealman RD, Risner ME, Henningfield JE (1983) Control of behavior by intravenous nicotine injections in laboratory animals. Pharmacol Biochem Behav 19:1011–1020

Grebenstein P, Burroughs D, Zhang Y, LeSage MG (2013) Sex differences in nicotine self-administration in rats during progressive unit dose reduction: implications for nicotine regulation policy. Pharmacol Biochem Behav 114–115:70–81

Guillem K, Vouillac C, Azar MR, Parsons LH, Koob GF, Cador M, Stinus L (2005) Monoamine oxidase inhibition dramatically increases the motivation to self-administer nicotine in rats. J Neurosci 25:8593–8600

Hall BJ, Wells C, Allenby C, Lin MY, Hao I, Marshall L, Levin ED (2014) Differential effects of non-nicotine tobacco constituent compounds on nicotine self-administration in rats. Pharmacol Biochem Behav 120:103–108

Harrison AA, Gasparini F, Marlou A (2002) Nicotine potentiation of brain stimulation reward reversed by DH beta E and SCH 23390, but not my eticlopride, LY 314582, or MPEP in rats. Psychopharmacology 160:56–66

Harrington GM (1963) Stimulus intensity stimulus satiation and optimum stimulation with light-contingent bar-press. Psychological Reports 13:107–111

Harvey DM, Yasar S, Heishman SJ, Panlilio LV, Henningfield JE, Goldberg SR (2004) Nicotine serves as an effective reinforcer of intravenous drug-taking behavior in human cigarette smokers. Psychopharmacology 175:134–142

Henningfield JE, Goldberg SR (1983) Nicotine as a reinforcer in human subjects and laboratory animals. Pharmacol Biochem Behav 19:989–992

Hoffman AC, Evans SE (2013) Abuse potential of non-nicotine tobacco smoke components: acetaldehyde, nornicotine, cotinine, and anabasine. Nicotine Tob Res 15:622–632

Houlgate PR, Dhingra KS, Nash SJ, Evans WH (1989) Determination of formaldehyde and acetaldehyde in mainstream cigarette smoke by high-performance liquid chromatography. Analyst 114:355–360

Huang H-Y, Hsieh S-H (2007) Analyses of tobacco alkaloids by cation-selective exhaustive injection sweeping microemulsion electrokinetic chromatography. J Chromatogr A 1164:313–319

Jensvold MF, Hamilton JA, Halbreich U (1996) Future research directions: methodological considerations for advancing gender-sensitive pharmacology. American Psychiatric Association, Arlington, VA, USA, pp 415–430

Johnson MW, Bickel WK, Kirshenbaum AP (2004) Substitutes for tobacco smoking: a behavioral economic analysis of nicotine gum, denicotinized cigarettes, and nicotine-containing cigarettes. Drug Alcohol Depend 74:253–264

Juliano LM, Donny EC, Houtsmuller EJ, Stitzer ML (2006) Experimental evidence for a causal relationship between smoking lapse and relapse. J Abnorm Psychol 115:166–173

Kenny PJ, Markou A (2006) Nicotine self-administration acutely activates brain reward systems and induces a long-lasting increase in reward sensitivity. Neuropsychopharmacology 31:1203–1211

Kim JYS, Fendrich M (2002) Gender differences in juvenile arrestees' drug use, self-reported dependence, and perceived need for treatment. Psychiatric Services 53:70–75

Kota D, Martin BR, Robinson SE, Damaj MI (2007) Nicotine dependence and reward differ between adolescent and adult male mice. J Pharmacol Exp Ther 322:399–407

Kotz D, Brown J, West R (2014) 'Real-world' effectiveness of smoking cessation treatments: a population study. Addiction 109:491–499

Kyerematen G, Owens G, Chattopadhyay B, Vesell E (1988) Sexual dimorphism of nicotine metabolism and distribution in the rat. Studies in vivo and in vitro. Drug Metab Dispos 16:823–828

Lang WJ, Latiff AA, McQueen A, Singer G (1977) Self administration of nicotine with and without a food delivery schedule. Pharmcol Biochem Behav 7:65–70

Lanza ST, Donny EC, Collins LM, Balster RL (2004) Analyzing the acquisition of drug self-administration using growth curve models. Drug Alcohol Depend 75:11–21

Lazev AB, Herzog TA, Brandon TH (1999) Classical conditions of environmental cues to cigarette smoking. Clinical Trial. Exp Clin Psychopharmacol 7:56–63

Le Foll B, Goldberg SR (2005) Control of the reinforcing effects of nicotine by associated environmental stimuli in animals and humans. Trends Pharmacol Sci 26:287–293

LeSage MG, Burroughs D, Dufek M, Keyler DE, Pentel PR (2004) Reinstatement of nicotine self-administration in rats by presentation of nicotine-paired stimuli, but not nicotine priming. Pharmacol Biochem Behav 79:507–513

Leslie FM, Mojica CY, Reynaga DD (2013) Nicotinic receptors in addiction pathways. Mol Pharmacol 83:753–758

Levin ED, Rezvani AH, Montoya D, Rose JE, Swartzwelder HS (2003) Adolescent-onset nicotine self-administration modeled in female rats. Psychopharmacology 169:141–149

Levin ED, Lawrence SS, Petro A, Horton K, Rezvani AH, Seidler FJ, Slotkin TA (2007) Adolescent vs. adult-onset nicotine self-administration in male rats: duration of effect and differential nicotinic receptor correlates. Neurotoxicol Teratol 29:458–465

Levin ED, Slade S, Wells C, Cauley M, Petro A, Vendittelli A, Johnson M, Williams P, Horton K, Rezvani AH (2011) Threshold of adulthood for the onset of nicotine self-administration in male and female rats. Behav Brain Res 225:473–481

Levin ME, Weaver MT, Palmatier MI, Caggiula AR, Sved AF, Donny EC (2012) Varenicline dose dependently enhances responding for nonpharmacological reinforcers and attenuates the reinforcement-enhancing effects of nicotine. Nicotine Tob Res 14:299–305

Li S, Zou S, Coen K, Funk D, Shram MJ, Le AD (2014) Sex differences in yohimbine-induced increases in the reinforcing efficacy of nicotine in adolescent rats. Addict Biol 19:156–164

Liu X, Caggiula AR, Yee SK, Nobuta H, Sved AF, Pechnick RN, Poland RE (2007) Mecamylamine attenuates cue-induced reinstatement of nicotine-seeking behavior in rats. Neuropsychopharmacology 32:710–718

Liu X, Caggiula AR, Palmatier MI, Donny EC, Sved AF (2008) Cue-induced reinstatement of nicotine-seeking behavior in rats: effect of bupropion, persistence over repeated tests, and its dependence on training dose. Psychopharmacology 196:365–375

Liu X, Jernigen C, Gharib M, Booth S, Caggiula AR, Sved AF (2010) Effects of dopamine antagonists on drug cue-induced reinstatement of nicotine-seeking behavior in rats. Behav Pharmacol 21:153–160

Loftipour Arnold SMM, Hogenkamp DJ, Gee KW, Belluzzi JD, Leslie FM (2011) The monoamine oxidase (MAO) inhibitor tranylcypromine enhances nicotine self-administration in rats through a mechanism independent of MAO inhibition. Neuropharm 61:95–104

Lynch WJ (2006) Sex differences in vulnerability to drug self-administration. Exp Clin Psychopharmacol 14:34–41

Lynch WJ (2009) Sex and ovarian hormones influence vulnerability and motivation for nicotine during adolescence in rats. Pharmacol Biochem Behav 94:43–50

Lynch WJ, Roth ME, Carroll ME (2002) Biological basis of sex differences in drug abuse: preclinical and clinical studies. Psychopharmacology 164:121–137

Markou A, Paterson NE (2009) Multiple motivational forces contribute to nicotine dependence. In: Bevins RA, Caggiula AR (eds) Nebraska Symposium on Motivation: The motivational impact of nicotine and its role in tobacco use, vol 55. Springer Science + Business Media, New York, pp 65–89

Matta SG, Balfour DJ, Benowitz NL, Boyd RT, Buccafusco JJ, Caggiula AR, Zirger JM (2007) Guidelines on nicotine dose selection for in vivo research. Psychopharmacology 190:269–319

McColl S, Sellers EM (2006) Research design strategies to evaluate the impact of formulations on abuse liability. Drug Alcohol Depend 83(Suppl. 1):S52–S62

McKee SA, Maciejewski PK, Falba T, Mazure CM (2003) Sex differences in the effects of stressful life events on changes in smoking status. Addiction 98:847–855

Melis M, Enrico P, Peana AT, Diana M (2007) Acetaldehyde mediates alcohol activation of the mesolimbic dopamine system. Eur J Neurosci 26:2824–2833

Mello NK, Fivel PA, Kohut SJ, Caine SB (2014) Anatabine significantly decreases nicotine self-administration. Exp Clin Psychopharmacol 22:1–8

Meyer PJ, Ma ST, Robinson TE (2011) A cocaine cue is more preferred and evokes more frequency-modulated 50 kHz ultrasonic vocalizations in rats prone to attribute incentive salience to a food cue. Psychopharmacology 219:999–1009

Myers WD, Ng KT, Singer G (1982) Intravenous self-administration of acetaldehyde in the rat as a function of schedule, food deprivation and photoperiod. Pharmacol Biochem Behav 17:807–811

Myers WD, Ng KT, Singer G (1984a) Effects of naloxone and buprenorphine on intravenous acetaldehyde self-injection in rats. Physiol Behav 33:449–455

Myers W, Ng K, Singer G (1984b) Ethanol preference in rats with a prior history of acetaldehyde self-administration. Cell Mol Life Sci 40:1008–1010

Natividad LA, Torres OV, Friedman TC, O'Dell LE (2013) Adolescence is a period of development characterized by short-and long-term vulnerability to the rewarding effects of nicotine and reduced sensitivity to the anorectic effects of this drug. Behav Brain Res 257:275–285

O'Dell LE, Torres OV (2014) A mechanistic hypothesis of the factors that enhance vulnerability to nicotine use in females. Neuropharmacology 76:566–580

O'Dell LE, Chen SA, Smith RT, Specio SE, Balster RL, Paterson NE, Koob GF (2007) Extended access to nicotine self-administration leads to dependence: Circadian measures, withdrawal measures, and extinction behavior in rats. J Pharmacol Exp Ther 320:180–193

O'Hara P, Portser SA, Anderson BP (1989) The influence of menstrual cycle changes on the tobacco withdrawal syndrome in women. Addict Behav 14:595–600

Palmatier MI, Evans-Martin FF, Hoffman A, Caggiula AR, Chaudhri N, Donny EC, Sved AF (2006) Dissociating the primary reinforcing and reinforcement-enhancing effects of nicotine using a rat self-administration paradigm with concurrently available drug and environmental reinforcers. Psychopharmacology 184:391–400

Palmatier MI, Liu X, Matteson GL, Donny EC, Caggiula AR, Sved AF (2007) Conditioned reinforcement in rats established with self-administered nicotine and enhanced by noncontingent nicotine. Psychopharmacology 195:235–243

Palmatier MI, Liu X, Donny EC, Caggiula AR, Sved AF (2008) Metabotropic glutamate five receptor (mGluR5) antagonists decrease nicotine seeking, but do not affect the reinforcement enhancing effects of nicotine. Neuropsychopharmacology 33:2139–2147

Palmatier MI, Levin ME, Mays KL, Donny EC, Caggiula AR, Sved AF (2009) Bupropion and nicotine enhance responding for nondrug reinforcers via dissociable pharmacological mechanisms in rats. Psychopharmacology 207:381–390

Palmatier MI, O'Brien LC, Hall MJ (2012) The role of conditioning history and reinforcer strength in the reinforcement enhancing effects of nicotine in rats. Psychopharmacology 219:1119–1131

Paterson NE, Markou A (2004) Prolonged nicotine dependence associated with extended access to nicotine self-administration in rats. Psychopharmacology 173:64–72

Pavlov IP (1927) Conditioned reflexes. An investigation of the physiological activity of the cerebral cortex. Oxford University Press, Oxford

Peana AT, Enrico P, Assaretti AR, Pulighe E, Muggironi G, Nieddu M, Piga A, Lintas A, Diana M (2008) Key role of ethanol-derived acetaldehyde in the motivational properties induced by intragastric ethanol: a conditioned place preference study in the rat. Alcohol Clin Exp Res 32:249–258

Peana AT, Muggironi G, Diana M (2010) Acetaldehyde-reinforcing effects: a study on oral self-administration behavior. Frontiers in Psychiatry 1:23

Peana AT, Muggironi G, Fois GR, Zinellu M, Vinci S, Acquas E (2011) Effect of opioid receptor blockade on acetaldehyde self-administration and ERK phosphorylation in the rat nucleus accumbens. Alcohol 45:773–783

Peartree NA, Sanabria F, Thiel KJ, Weber SM, Cheung TH, Neisewander JL (2012) A new criterion for acquisition of nicotine self-administration in rats. Drug Alcohol Depend 124:63–69

Perkins KA (2009) Does smoking cue-induced craving tell us anything important about nicotine dependence? Addiction 104:1610–1616

Perkins KA (2009) Sex differences in nicotine reinforcement and reward: influences on the persistence of tobacco smoking. Nebr Symp Motiv 55:143–169

Perkins KA, Sexton JE, Reynolds WA, Grobe JE, Fonte C, Stiller RL (1994) Comparison of acute subjective and heart rate effects of nicotine intake via tobacco smoking versus nasal spray. Pharmacol Biochem Behav 47:295–299

Perkins KA, Karelitz JL (2013a) Influence of reinforcer magnitude and nicotine amount on smoking's acute reinforcement enhancing effects. Drug Alcohol Depend 133:167–171

Perkins KA, Karelitz JL (2013b) Reinforcement enhancing effects of nicotine via smoking. Psychopharmacology 228:479–486

Perkins KA, Donny E, Caggiula AR (1999) Sex differences in nicotine effects and self-administration: review of human and animal evidence. Nicotine Tob Res 1:301–315

Perkins KA, Gerlach D, Vender J, Meeker J, Hutchison S, Grobe J (2001) Sex differences in the subjective and reinforcing effects of visual and olfactory cigarette smoke stimuli. Nicotine Tob Res 3:141–150

Perkins KA, Jacobs L, Sanders M, Caggiula AR (2002) Sex differences in the subjective and reinforcing effects of cigarette nicotine dose. Psychopharmacology 163:194–201

Perkins KA, Karelitz JL, Giedgowd GE, Conklin CA (2013) Negative mood effects on craving to smoke in women versus men. Addict Behav 38:1527–1531

Picciotto MR, Caldarone BJ, Brunzell DH, Zachariou V, Stevens TR, King SL (2001) Neuronal nicotinic acetylcholine receptor subunit knockout mice: physiological and behavioral phenotypes and possible clinical implications. Pharmacol Ther 92:89–108

Quertemont E, De Witte P (2001) Conditioned stimulus preference after acetaldehyde but not ethanol injections. Pharmacol Biochem Behav 68:449–454

Rescorla RA, Solomon RL (1967) Two-process learning theory: relationships between Pavlovian conditioning and instrumental learning. Psychol Rev 74:151–182

Robbins TW, Koob GF (1978) Pipradrol enhances reinforcing properties of stimuli paired with brain stimulation. Pharmacol Biochem Behav 8:219–222

Robinson TE, Berridge KC (2000) The psychology and neurobiology of addiction: an incentive-sensitization view. Addiction 95:S91–S117

Robinson TE, Yager LM, Cogan ES, Saunders BT (2014) On the motivational properties of reward cues: individual differences. Neuropharmacology 76:450–459

Rodd-Henricks ZA, Melendez RI, Zaffaroni A, Goldstein A, McBride WJ, Li T-K (2002) The reinforcing effects of acetaldehyde in the posterior ventral tegmental area of alcohol-preferring rats. Pharmacol Biochem Behav 72:55–64

Rose JE, Behm FM, Westman EC, Coleman RE (1999) Arterial nicotine kinetics during cigarette smoking and intravenous nicotine administration: implications for addiction. Drug Alcohol Depend 56:99–107

Rose JE, Behm FM, Westman EC, Johnson M (2000) Dissociating nicotine and nonnicotine components of cigarette smoking. Pharmacol Biochem Behav 67:71–81

Russell MAH, Feyerabend C (1978) Cigarette smoking: a dependence on high-nicotine boli. Drug Metab Rev 8:29–57

Russell JC, Epling WF, Pierce D, Amy RM, Boer DP (1987) Induction of voluntary prolonged running by rats. J Appl Physiol 63:2549–2553

Samaha A-N, Robinson TE (2005) Why does the rapid delivery of drugs to the brain promote addiction? Trends Pharmacol Sci 26:82–87

Saunders BT, Robinson TE (2011) Individual variation in the motivational properties of cocaine. Neuropsychopharmacology 36:1668–1676

Sayette MA, Tiffany ST (2013) Peak-provoked craving deserves a seat at the research table. Addiction 108:1030–1031

Schassburger RL, Rupprecht LE, Smith TT, Buffalari DM, Thiels E, Donny EC, Sved AF (2013) Nicotine enhances the rewarding properties of sucrose. Paper presented at the Society for Neuroscience, San Diego, California

Schnoll RA, Patterson F, Lerman C (2007) Treating tobacco dependence in women. J Women's Health 16:1211–1218

Shram MJ, Funk D, Li Z, Lê AD (2008a) Nicotine self-administration, extinction responding and reinstatement in adolescent and adult male rats: evidence against a biological vulnerability to nicotine addiction during adolescence. Neuropsychopharmacology 33:739–748

Shram MJ, Li Z, Lê AD (2008b) Age differences in the spontaneous acquisition of nicotine self-administration in male Wistar and Long-Evans rats. Psychopharmacology 197:45–58

Siegel S (1988) Drug anticipation and the treatment of dependence. NIDA Res Monogr 84:1–24

Singer G, Simpson F, Lang WJ (1978) Schedule induced self injections of nicotine with recovered body weight. Pharmacol Biochem Behav 9:387–389

Skinner BF (1953) Science and Human Behavior. Macmillan, New York

Smith B, Amit Z, Splawinsky J (1984) Conditioned place preference induced by intraventricular infusions of acetaldehyde. Alcohol 1:193–195

Smith TT, Levin ME, Schassburger RL, Buffalari DM, Sved AF, Donny EC (2013) Gradual and immediate nicotine reduction result in similar low-dose nicotine self-administration. Nicotine Tob Res 15:1918–1925

Smith TT, Schassburgher RL, Rupprecht LE, Buffalari DM, Sved AF, Donny EC (2014a) Effects of tranylcypromine, an irreversible monoamine oxidase (MAO) inhibitor, on nicotine self-administration in rats. Paper presentation at the annual meeting of the association for behavior analysis international, Chicago, Illinois

Smith TT, Schassburger RL, Buffalari DM, Sved AF, Donny EC (2014b) Low dose nicotine self-administration is reduced in adult male rats naïve to high doses of nicotine: implications for nicotine product standards. Exp Clin Psychopharmacol. http://dx.doi.org/10.1037/a0037396

Sofuoglu M, Yoo S, Hill KP, Mooney M (2008) Self-administration of intravenous nicotine in male and female cigarette smokers. Neuropsychopharmacology 33:715–720

Sorge RE, Clarke PB (2009) Rats self-administer intravenous nicotine delivered in a novel smoking-relevant procedure: effects of dopamine antagonists. J Pharmacol Exp Ther 330:633–640

Sorge RE, Pierre VJ, Clarke PB (2009) Facilitation of intravenous nicotine self-administration in rats by a motivationally neutral sensory stimulus. Psychopharmacology 207:191–200

Speakman JR, Mitchel SE (2011) Caloric Restriction. Mol Aspects Med 32:159–221

Spear LP (2000) The adolescent brain and age-related behavioral manifestations. Neurosci Biobehav Rev 24:417–463

Spina L, Longoni R, Vinci S, Ibba F, Peana AT, Muggironi G, Spiga S, Acquas E (2010) Role of dopamine D1 receptors and extracellular signal regulated kinase in the motivational properties of acetaldehyde as assessed by place preference conditioning. Alcohol Clin Exp Res 34:607–616

Stolerman IP (1989) Discriminative stimulus effects of nicotine in rats trained under different schedules of reinforcement. Psychopharmacology 97:131–138

Stolerman IP (1999) Inter-species consistency in the behavioural pharmacology of nicotine dependence. Behav Pharmacol 10:559–580

Stolerman IP, Jarvis M (1995) The scientific case that nicotine is addictive. Psychopharmacology 117:2–10

Takayama S, Uyeno E (1985) Intravenous self-administration of ethanol and acetaldehyde by rats. Japan J Psychopharmacol 5:329–334

Tiffany ST (1990) A cognitive model of drug urges and drug-use behavior: role of automatic and nonautomatic processes. Psychol Rev 97:147–168

Torres OV, Tejeda HA, Natividad LA, O'Dell LE (2008) Enhanced vulnerability to the rewarding effects of nicotine during the adolescent period of development. Pharmacol Biochem Behav 90:658–663

USDHHS (1988) Office of the Surgeon General DHHS Publication no. (CDC): 88-8406.0. The health consequences of smoking: nicotine addiction: a report of the surgeon general. Center for Health Promotion and Education. Office on Smoking and Health United States. Public Health Service

USDHHS (2012) Preventing tobacco use among youth and young adults: a report of the Surgeon General. US Department of Health and Human Services, Centers for Disease Control and Prevention, National Center for Chronic Disease Prevention and Health Promotion, Office on Smoking and Health, Atlanta, GA, 3

Vastola BJ, Douglas LA, Varlinskaya EI, Spear LP (2002) Nicotine-induced conditioned place preference in adolescent and adult rats. Physiol Behav 77:107–114

Villegier AS, Lotfipour S, Belluzzi JD, Leslie FM (2007a) Involvement of alpha1-adrenergic receptors in tranylcypromine enhancement of nicotine self-administration in rat. Psychopharmacology 193:457–465

Villegier AS, Lotfipour S, McQuown SC, Belluzzi JD, Leslie FM (2007b) Tranylcypromine enhancement of nicotine self-administration. Neuropharmacology 52:1415–1425

Wakasa Y, Takada K, Yanagita T (1995) Reinforcing effect as a function of infusion speed in intravenous self-administration of nicotine in rhesus monkeys. Japan J Psychopharmacol 15:53–59

Wertz JM, Sayette MA (2001) A review of the effects of perceived drug use opportunity of self-reported urge. Exp Clin Psychopharmacol 9:3–13

Wing VC, Shoaib M (2008) Contextual stimuli modulate extinction and reinstatement in rodents self-administering intravenous nicotine. Psychopharmacology 200:357–365

Wing VC, Shoaib M (2013) Effect of infusion rate on intravenous nicotine self-administration in rats. Behav Pharmacol 24:517–522

Wray JM, Godleski SA, Tiffany ST (2011) Cue-reactivity in the natural environment of cigarette smokers: the impact of photographic and in vivo smoking stimuli. Psychol Addict Behav 25:733–737

Xie J, Yin J, Sun S, Xie F, Zhang X, Guo Y (2009) Extraction and derivatization in single drop coupled to MALDI-FTICR-MS for selective determination of small molecule aldehydes in single puff smoke. Anal Chim Acta 638:198–201

Xu J, Azizian A, Monterosso J, Domier CP, Brody AL, London ED, Fong TW (2008) Gender effects on mood and cigarette craving during early abstinence and resumption of smoking. Nicotine Tob Res 10:1653–1661

Yager LM, Robinson TE (2010) Cue-induced reinstatement of food seeking in rats that differ in their propensity to attribute incentive salience to food cues. Behav Brain Res 214:30–34

Yan Y, Pushparaj A, Gamaleddin I, Steiner RC, Picciotto MR, Roder J, Le Foll B (2012) Nicotine-taking and nicotine-seeking in C57Bl/6 J mice without prior operant training or food restriction. Behav Brain Res 230:34–39

Zou S, Funk D, Shram MJ, Le AD (2014) Effects of stressors on the reinforcing efficacy of nicotine in adolescent and adult rats. Psychopharmacology 231:1601–1614

The Role of Mesoaccumbens Dopamine in Nicotine Dependence

David J.K. Balfour

Abstract There is abundant evidence that the dopamine (DA) neurons that project to the nucleus accumbens play a central role in neurobiological mechanisms underpinning drug dependence. This chapter considers the ways in which these projections facilitate the addiction to nicotine and tobacco. It focuses on the complimentary roles of the two principal subdivisions of the nucleus accumbens, the accumbal core and shell, in the acquisition and maintenance of nicotine-seeking behavior. The ways in which tonic and phasic firing of the neurons contributes to the ways in which the accumbens mediate the behavioral responses to nicotine are also considered. Experimental studies suggest that nicotine has relatively weak addictive properties which are insufficient to explain the powerful addictive properties of tobacco smoke. This chapter discusses hypotheses that seek to explain this conundrum. They implicate both discrete sensory stimuli closely paired with the delivery of tobacco smoke and contextual stimuli habitually associated with the delivery of the drug. The mechanisms by which each type of stimulus influence tobacco dependence are hypothesized to depend upon the increased DA release and overflow, respectively, in the two subdivisions of the accumbens. It is suggested that a majority of pharmacotherapies for tobacco dependence are not more successful because they fail to address this important aspect of the dependence.

Keywords Dopamine · Extracellular · Nucleus accumbens shell · Nucleus accumbens core · Conditioned stimuli · Contextual conditioning

D.J.K. Balfour (✉)
Medical Research Institute, Division of Neuroscience, Ninewells Hospital and Medical School, Dundee DD1 9SY, Scotland
e-mail: d.j.k.balfour@dundee.ac.uk

D.J.K. Balfour and M.R. Munafò (eds.), *The Neuropharmacology of Nicotine Dependence*, Current Topics in Behavioral Neurosciences 24,
DOI 10.1007/978-3-319-13482-6_3

Contents

1 General Introduction

For the last 20 years or so it has been an axiom of research in the field of drug dependence that most, if not all, addictive drugs stimulate dopamine (DA) release in the limbic areas of the forebrain, especially the nucleus accumbens and that this property of the drugs plays a central role in their ability to cause dependence (Dackis and Gold 1985; Di Chiara 1999; Di Chiara and Imperato 1988; Gold and Dackis 1984; Wise 1987; Wise and Bozarth 1987). Early theories suggested that increased DA release in the nucleus accumbens mediated the powerful rewarding or euphoriant properties of the drugs and that this underpinned the development of compulsive drug-seeking and drug-taking behaviors (Wise 1988a, b, 1996; Wise and Bozarth 1985). The most recent work, however, suggests that the DA projections to the forebrain play a more complex role in drug dependence (Balfour 2009; Di Chiara and Bassareo 2007; Volkow et al. 2011; Willuhn et al. 2010).

Self-administration experiments have long been seen as the gold standard for exploring neurobiological mechanisms that mediate the reinforcing or rewarding properties of addictive drugs (see chapter entitled Behavioral Mechanisms Underlying Nicotine Reinforcement; this volume). Studies using this model of reinforcement have shown that selective lesions of the DA projections to the nucleus accumbens markedly attenuate acquisition of responding for a range of addictive drugs, including cocaine, amphetamine and heroin, with different molecular targets or mechanisms of action (Balfour 2009; Di Chiara and Bassareo 2007; Lyness et al. 1979; Roberts et al. 1977; Roberts and Koob 1982; Singer and Wallace 1984; Volkow et al. 2011; Willuhn et al. 2010). These results point to a pivotal role for this pathway in drug reward and reinforcement. Early experiments showed that the effects of the lesions on these DA projections also attenuate nicotine self-administration in rats (Singer et al. 1982). Subsequent studies by Corrigall and Coen (1989) established a more robust experimental design for the investigation of nicotine self-administration in rats and used this procedure to confirm that lesions of the pathway do, indeed, attenuate nicotine self-administration in this species (Corrigall et al. 1992). These results provide strong support for the hypothesis that nicotine shares at least some of the key properties of a drug abuse and that DA projections from the ventral tegmental area (VTA) play a key role in these responses. This chapter will review the role of DA in nicotine dependence and seek to address the complex nature of the relationship between behavioral measures of nicotine dependence and its effects on DA release in the nucleus accumbens.

There is evidence that stimulation DA neurons can result in the release of DA both into synapses within the nucleus accumbens and into the interstitial space that lies between the fibers and that DA release into these two compartments has differing roles in the regulation of behavior (Floresco 2007; Grace 1991; Heien and Wightman 2006). Figure 1 highlights the complexity of the pathways projecting from within the VTA and areas such as the nucleus accumbens, the frontal cortex, ventral pallidum, pedunculopontine/laterodorsal tegmentum, and pedunculopontine and laterodorsal tegmentum, pedunculopontine and laterodorsal tegmentum that influence the activity of the DA projections from the VTA (Vanderschuren and Kalivas 2000; Sesack and Grace 2010). DA release in the nucleus accumbens is also influenced by projections to the accumbens from the ventral subiculum of the hippocampus, the basolateral amygdala and the frontal cortex. These complex mechanisms for controlling DA release in the nucleus accumbens allow the system to respond to external and internal contingencies and stimuli which predict reward and, thus, drive positively rewarded goal-directed behaviors. It has been proposed that chronic exposure to most drugs of abuse results in impaired regulation of these neurons in a manner that results in the inappropriate behavioral responses and strategies that characterize dependence (Floresco 2007; Grace 1991; Grace et al. 2007; Heien and Wightman 2006; Schultz 2010, 2011; Sesack and Grace 2010).

2 The Effects of Nicotine on DA Release from Mesolimbic DA Neurons In Vivo

There is considerable evidence from electrophysiological and microdialysis studies that nicotine falls into the category of addictive drugs which stimulate DA release in the nucleus accumbens. Studies in the mid-1980 showed that injections of nicotine stimulate midbrain DA neurons (Grenhoff et al. 1986). Imperato et al. (1986) were among the first to report that nicotine injections stimulate DA overflow in the

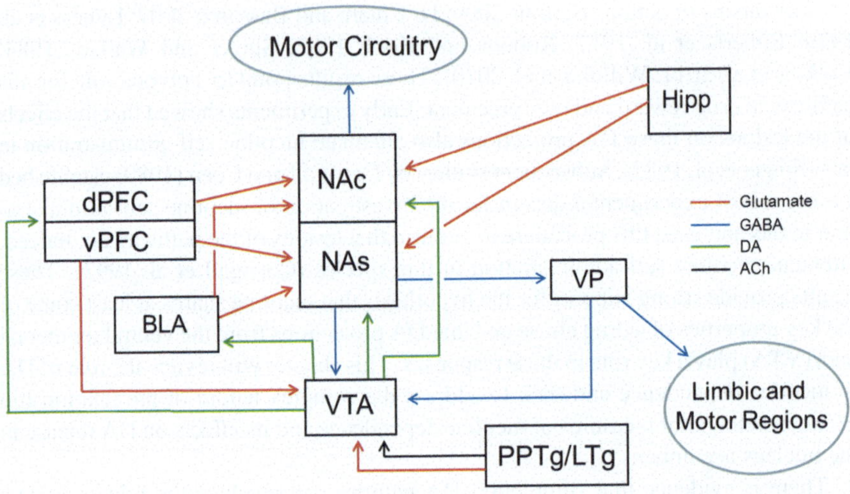

Fig. 1 Innervation of the nucleus accumbens. This figure highlights the principal pathways that innervate and project from the nucleus accumbens core and shell. *NACs* nucleus accumbens shell; *NAcC* nucleus accumbens core; *VTA* ventral tegmental area; *dPFC* dorsal prefrontal cortex; *cPFC* ventral prefrontal cortex; *VP* ventral pallidum; *HIPP* hippocampus; *PPTg* pedunculopontine tegmental nuclei; *LTg* lateral tegmental nuclei. The figure is adapted from Vanderschuren and Kalivas (2000)

nucleus accumbens when measured using microdialysis. In the years following, other studies confirmed these initial observations (Benwell and Balfour 1992; Brazell et al. 1990; Damsma et al. 1989; Mereu et al. 1987; Nisell et al. 1994b; Schilstrom et al. 2003). The nucleus accumbens is composed of two principal subdivisions, the accumbal shell and the accumbal core, which are anatomically distinct, project to different areas of the brain and are thought to subserve different physiological and behavioral functions (Heimer et al. 1991; Zahm and Brog 1992). Acute injections of drugs of dependence elicit greater increases in DA overflow in the shell subdivision of the accumbens that the accumbal core (Di Chiara and Bassareo 2007). The acute administration of nicotine to drug-naïve animals also results in a preferential increase in DA overflow in the accumbal shell (Cadoni and Di Chiara 2000). Thus, the effects of acute nicotine on DA overflow in the nucleus accumbens closely resemble those observed for other drugs of dependence (Balfour 2004; Di Chiara 2000; Pontieri et al. 1996).

The DA neurons which project from the VTA express neuronal nicotinic receptors on both their somatodendric membranes and the membranes of the nerve terminals and peri-terminal regions of the axons (Livingstone and Wonnacott 2009). The molecular composition of these receptors is diverse and is considered in some detail by Fasoli and Gotti (see chapter entitled Structure of Neuronal Nicotinic Receptors; this volume). However, there is compelling evidence that the effects of systemic nicotine administration to rats depend upon the stimulation of one or more of these receptor populations (Livingstone and Wonnacott 2009). It is also important to remember that a number of the neuronal pathways which project to the VTA or the nucleus accumbens and influence the activity of DA neurons also express neuronal nicotinic receptors, and it is widely accepted that stimulation of these receptors also contributes significantly to the effects of systemic nicotine on DA release in the terminal fields of the mesolimbic system (Markou 2008). The effects of nicotine on DA overflow in the nucleus accumbens are attenuated by the prior administration of nicotinic receptor antagonists that penetrate into the brain (Benwell et al. 1995). Furthermore, the local administration of a nicotinic receptor antagonist into the VTA, but not the nucleus accumbens, also attenuates the increase in DA overflow evoked by a systemic injection of nicotine, results which support the conclusion that the changes in DA overflow evoked by nicotine depend upon the stimulation of nicotinic receptors in the VTA resulting in increase impulse flow to the terminal field (Nisell et al. 1994a, b). Additional studies suggest that the reinforcing properties of nicotine, measured using a self-administration paradigm, also depend upon the stimulation of the DA neurons in the VTA which project to the accumbens (Corrigall et al. 1994). These results support the conclusion that effects of self-administered nicotine on DA release in the forebrain, which underpin nicotine reinforcement, depend upon its ability to stimulate impulse flow to the terminal fields.

A single acute injection of nicotine elicits a sustained increase in the concentration of DA in the extracellular space of the accumbens shell that persists for up to 60 min. However, there is evidence that many of receptors on the DA neurons in the VTA rapidly desensitize upon the exposure to the drug (Pidoplichko et al. 1997). In order to explain this conundrum, Mansvelder and McGehee (Mansvelder et al. 2002; Mansvelder and McGehee 2000, 2002) suggested that sustained increases in firing of the neurons were maintained by the differential desensitization of the nicotinic receptors located the DA neurons and the terminals of afferent GABA and glutamate neurons that regulate the activity of the neurons. They argued that this generates a triphasic response to the drug. When nicotine is administered, it initially stimulates the nicotinic receptors located on the somatodendritic membranes of the DA neurons. This leads to an initial short-lived stimulation of the neurons. Nicotine also stimulates nicotinic receptors on GABA terminals facilitating the release of this inhibitory transmitter. This results a short-lived inhibition of the DA neurons following the initial stimulation. These authors suggested that receptors on the DA neurons and GABA terminals are composed of $\alpha4\beta2$ subunits and readily desensitize within seconds or minutes of exposure to the drug. A third population of nicotinic receptors, located on glutamate terminals, is composed predominantly of $\alpha7$ subunits. These receptors are more resistant to desensitization and remain active over a longer period of time, facilitating the release of glutamate in the VTA. The authors suggest that the sustained increase in neuronal activity and DA overflow, evoked by the acute administration of nicotine, depends upon this sustained release of glutamate and that this mechanism is important to the role of DA in nicotine dependence (Fagen et al. 2003).

The paragraphs above have focused on the DA responses to acute nicotine administered to drug-naïve animals. These studies have generated convincing evidence that nicotine injections stimulate the mesoaccumbens DA neurons that have been implicated in drug dependence. However, the development of nicotine dependence is the product of chronic or repeated exposure to the drug, and if we are to understand the role of these projections in dependence, then it is essential that the changes in the neural responses to nicotine that occur with chronic exposure to the drug are fully characterized. Studies in our laboratory were the first to explore the effects of repeated daily injections of nicotine on DA overflow in the nucleus accumbens (Benwell and Balfour 1992). These experiments showed that, whereas acute injections of nicotine had little or no effects on DA overflow in the core subdivision of the nucleus accumbens, repeated administration of the drug, on as few as five occasions, was sufficient to elicit sensitization of its effects on DA overflow in this region of the brain. Some years later, this observation was confirmed by Cadoni and Di Chiara (2000) who also demonstrated that sensitization was only observed in the core subdivision of the accumbens. Repeated administration of nicotine does not result in sensitization of the DA overflow evoked by nicotine in the accumbal shell. Indeed, in this subdivision of the nucleus accumbens, repeated nicotine administration results in the development of tolerance to the response to nicotine (Cadoni and Di Chiara 2000), an observation subsequently confirmed in our own laboratory (Iyaniwura et al. 2001; Tronci and Balfour 2011). Furthermore, repeated nicotine administration does not result in sensitization of DA overflow in the dorsolateral striatum (Benwell and Balfour 1997). Repeated daily administration of other drugs of abuse also results in a regionally selective sensitization of their effects of DA overflow in the core subdivision of the accumbens (Cadoni and Di Chiara 1999; Cadoni et al. 2000).

The extent to which the electrophysiological responses to nicotine are modified by chronic or repeated exposure to the drug remains to be clarified. However, there is evidence that pretreatment with nicotine diminishes the regulation by D2 autoreceptors of DA release in the nucleus accumbens (Balfour et al. 1998). By contrast, nicotine pretreatment does not alter the regulatory effects of these receptors on DA overflow in the dorsoalateral striatum, data which suggest that the regionally selective sensitization of the response to nicotine in the accumbal core could be associated with diminished autoreceptor regulation of the neurons. Other studies in our laboratory demonstrated some years ago that both the development and expression of the sensitized DA responses to nicotine, observed in the accumbens core, are inhibited by pretreating the animals with an NMDA receptor antagonist (Balfour et al. 1996; Shoaib et al. 1994). These results imply that the enhanced overflow of DA in the accumbal core depends upon the co-stimulation of the glutamatergic receptors. The anatomical location of the receptors that mediate the sensitized responses to nicotine has not been identified although some data support the conclusion that the receptors are likely to be located in the VTA and to modulate the effects of nicotine on impulse flow to the terminal field (Fu et al. 2000; Schilstrom et al. 1998). This conclusion is consistent with the evidence that the sustained increases in DA overflow, evoked by a nicotine injection, are maintained by the release of glutamate in the VTA (Mansvelder et al. 2002). The sensitized DA response is also attenuated by pretreatment with the indole alkaloid, ibogaine, an effect that persists for at least 24 h after administration of the alkaloid (Benwell et al. 1996).

The compound has been reported to attenuate the increases in DA overflow evoked by other drugs of dependence, and this property is thought to underpin the potential of this compound and its synthetic derivative 18-methoxycoronaridine as treatments for drug dependence (Glick and Maisonneuve 2000). Ibogaine is an antagonist at α4β2 nicotinic receptors, and its effects at this receptor are long lasting. Thus, this could explain the inhibition evoked by the drug. However, if this is the mechanism, it is surprising that pretreatment with ibogaine does not also attenuate the increase in locomotion evoked by nicotine. Ibogaine is also an NMDA receptor antagonist, and since its effects on the responses to nicotine resemble more closely those of other NMDA antagonists, it seems more likely that its effects at this receptor explain the interaction with nicotine in our experiments (Benwell et al. 1996).

For a majority of the studies described in the previous paragraph, DA overflow was measured in the absence of an inhibitor of the neuronal DA transporter. It is assumed, therefore, that the measures reflect changes in the extracellular DA concentration that occur in vivo in response to the administration of the drug. However, the electrophysiological studies suggest that the effects of nicotine on the neuronal firing are not restricted to the neurons that project to the accumbal shell. Some years ago, our group investigated the effects of nicotine on catecholamine overflow measured in the presence of a drug, nomifensine, which inhibits the neuronal DA and noradrenaline transporters (Benwell and Balfour 1997). In the presence of this inhibitor, acute injections of nicotine to drug-naïve animals also increase DA overflow in the accumbal core and dorsolateral striatum, whereas no changes are observed if the experiment is performed in the absence of the inhibitor. These data imply that the acute systemic administration of nicotine exerts a more generalized stimulation of the activity of DA neurons in the VTA and the substantia nigra but that the effects are masked by the efficient reuptake of the monoamine by the transporters. The conclusion is supported by more recent studies which suggest that the increase in DA release evoked by changes in phasic burst firing of the neurons can be detected in the presence of an inhibitor of the transporters (Floresco 2007). Thus, it is possible that the regionally selective increase in DA overflow in the accumbal shell, measured in response to acute nicotine administration, could reflect differences in nature of the effects of the drug on the activity of the DA neurons that project to this subdivision of the accumbens. Jones et al. (1996), however, have highlighted evidence that the characteristics and regulation of the DA transporters in the accumbal shell differ from those found in the core subdivision of the structure and that this may explain the regional differences in the DA responses in the accumbens.

3 Tonic and Phasic Firing of Mesolimbic Dopamine Neurons

DA neurons have been shown to fire in two distinct modes—as irregular single spikes and as short periods of burst firing (Grace and Bunney 1984a, b). Stimulation of the burst firing of the neurons evokes a substantial increase in the DA release into the synaptic cleft and generally leads to only a transitory increase in extracellular DA (Heien and Wightman 2006). By contrast, the concentration of DA in the extracellular space is governed primarily by the tonic spike activity of the neurons (Floresco 2007; Floresco et al. 2003; Heien and Wightman 2006) although a recent study by Owesson-White et al. (2012) suggests that transient phasic firing of the neurons may also contribute to changes in the concentration of DA in the extracellular compartment of the nucleus accumbens. Nirenberg et al. (1997) used high-resolution electron microscopic immunocytochemistry to demonstrate that a majority of the neuronal DA transporters are located extra-synaptically on plasma membranes of both the accumbens shell and core. DA transporters are similarly predominantly located outside synapses in the dorsal striatum (Hersch et al. 1997). These observations support the conclusion, first mooted by Garris and Wightman (1994), that the primary role of DA transporters in the nucleus accumbens is to regulate the concentration of DA in the extracellular space. The release of DA into the extracellular space could reflect overflow from synapses. However, Nirenberg et al. (1997) noted that the DA fibers that project into the nucleus accumbens are beaded, having varicosities along their length. This observation led Balfour et al. (2000) to propose that tonic DA release may reflect release directly into the extracellular space from these varicosities and that this DA serves a different functional role from that released by terminals that form tight synaptic contacts.

An important putative role for extracellular DA in the VTA may be autoreceptor-mediated inhibition of DA release evoked by phasic firing of the neurons (Grace 2000). Thus, the level of DA tone may influence the level of response to conditioned stimuli whose salience has been established by pairing with a rewarding outcome. There is also evidence that DA released into the extracellular compartment may be implicated in volume transmission, a form of neurotransmission that allows the transmitter to elicit a tonic regulation of "hard-wired" neuronal pathways which may be close to or some distance from the DA neurons (Fuxe et al. 2010; Garris and Wightman 1994). In this way, the transmitter may serve a "paracrine-like" role as a gate which influences the intensity of the neural traffic within fibers of passage. This possibility is consistent with the role attributed to the DA projections to the accumbens suggested by Sesack and Grace (2010). This chapter will focus on the possibility that DA release into the extracellular space of the accumbens and dorsal striatum may play complementary roles in the development of nicotine dependence.

There is also a considerable and growing body of evidence to suggest that the different modes of firing of the DA neurons which project to the nucleus accumbens mediate different components of the behavioral responses which depend upon the increased DA release in this area of the brain. Under normal conditions, many of the DA-secreting neurons in the VTA are inhibited by very active GABA neurons which project from a number of anatomical sites. The GABA projections from the ventral pallidum seem to play a particularly important role in suppressing the spontaneous activity of mesencephalic DA neurons (Floresco et al. 2001, 2003). As a result, approaching 50 % of the DA neurons in the VTA is not spontaneously active. Inactivation of the ventral pallidum releases this inhibitory effect and allows more of the neurons to fire spontaneously (Floresco et al. 2003; Lodge and Grace 2006a). Furthermore, excitation of glutamate afferents in the ventral subiculum of the hippocampus also inhibits the GABA projections from the ventral pallidum to the VTA by stimulating inhibitory GABA projections from the nucleus accumbens to the ventral pallidum. As a result, the DA neurons within the VTA are disinhibited, resulting in an increase in the proportion of the cells which are spontaneously active (Grace et al. 2007; Lodge and Grace 2006a). Phasic burst firing of the neurons is promoted by stimulation of NMDA receptors located on the cells (Chergui et al. 1993). Stimulation of these receptors is not alone sufficient to generate burst firing of the neurons. Burst firing is generated by the coincidental stimulation of the NMDA receptors and other stimulatory receptors located on the neurons. Other studies suggest that co-stimulation of a projection from the laterodorsal tegmentum is also required to elicit burst firing of the neurons (Lodge and Grace 2006b). The nature of the neurotransmitter released by the neurons that project from the LTDg remains unclear although some evidence suggests that the neurons may be cholinergic in nature and that the acetylcholine released by these neurons acts on nicotinic receptors putatively located on the DA cells (Mameli-Engvall et al. 2006). Thus, the LDTg seems to provide a cholinergic-dependent permissive "gate" that allows stimulation of the glutamate receptors to result in the expression of burst firing. The source of the glutamatergic pathways which elicit burst firing also remains to be established with certainty although there is evidence that glutamate projections from the pedunculopontine tegmentum (PPTg) directly regulate burst firing of DA neurons in the VTA. The PPTg receives projections for other sites within the brain (e.g., the prefrontal cortex and extended amygdala) and is, therefore, ideally suited to the integration of the effects of a range of sensory inputs on the burst firing of VTA DA neurons (Grace et al. 2007).

Selective ablation of the cholinergic neurons in the PPTg reduces nicotine self-administration (Lanca et al. 2000). Furthermore, this group also showed that microinjections of the β2 subunit-selective nicotinic receptor antagonist, dihydro-β-erythroidine, into the PPTg exert a similar effect on nicotine self-administration. The data, when taken together, support the hypothesis that stimulation of cholinergic neurons within the PPTg is required to stimulate the DA neurons in the VTA and, thus, mediate the reinforcing properties of nicotine. Subsequent studies suggested that various pharmacological manipulations of neurons within the PPTg exert common effects on nicotine and cocaine self-administration although GABAergic neurons within the structure appear to exert a selective effect on the self-administration of nicotine (Corrigall et al. 2001, 2002). A later study by Alderson et al. (2006) found that lesioning the posterior PPTg with the excitotoxin, ibotenic acid, increased nicotine self-administration. By contrast, similar lesions of the anterior PPTg were without effect. The significance of these findings lies in the fact that the neurons in the posterior PPTg project to the VTA, whereas those in the anterior PPTg project preferentially to the substantia nigra. The effects of the posterior PPTg lesions on nicotine self-administration were hypothesized to reflect either a reduction in intrinsic nicotine reward value or enhancement of associative incentive salience. The results, however, do not preclude the possibility that the lesions ablate GABA neurons in the posterior PPTg that are implicated in the regulation of DA neurons in the VTA.

The phasic activity of mesencephalic DA neurons is stimulated by exposure to external stimuli and signals the relationship between an external stimulus and the outcome of a behavioral response. The system is most effective at encoding information about rewards although some DA neurons within the VTA also respond to aversive outcomes and may encode information about stimuli associated with these outcomes (Schultz 2010). Phasic firing of the neurons is initially stimulated by exposure to an unanticipated reward (i.e., when the presentation of a stimulus first results in the presentation of a reward such as a sweetened pellet). As the animal learns the association, the stimulus takes on the properties of a conditioned stimulus and the presentation of the stimulus results in phasic activity rather than the presentation of the reward. A failure to present the anticipated reward results in a depression of phasic activity, an outcome described as a prediction error. The magnitude of the phasic response is influenced by the magnitude of the reward presented and by the length of the delay between the stimulus and the presentation of the reward. Thus, the degree of phasic firing seems to code for the subjective value of the reward rather than its objective value (Schultz 2010). If the presentation of the conditioned stimulus prompts a behavioral response (e.g., a lever press) that delivers the reward, then it can be seen that an important role of phasic firing of the neurons is to promote a behavioral response which is advantageous to the animal. Schultz has proposed that this process facilities economic decision making—the selection of the best response to the stimuli presented (Schultz 2007, 2010). Schultz (2011) argues that chronic exposure the drugs of abuse elicits changes in the regulation of midbrain DA neurons and postsynaptic DA receptors and the morphology of DA-receptive cells that impair the ability of the individual to discriminate between rewards. Furthermore, chronic exposure to the drugs increases the effects of prediction errors (when a reward predicted by a stimulus is not presented) and enhance neuronal responses to reward-predicting stimuli.

3.1 The Effects of Nicotine on Phasic and Tonic Firing of Dopamine Neurons

Early studies suggested that the acute injections of nicotine to experimental animals increased both the population spike activity and burst firing of the DA neurons that project from the both the substantia nigra and VTA (Grenhoff et al. 1986). More recent studies have shown that the administration nicotine to experimental animals predominantly increases the number, duration, and frequency of phasic burst firing events evoked in the DA neurons of the VTA and substantia nigra (Zhang et al. 2009). These authors also found that there are fundamental differences in the frequency-dependent release of DA from terminals in the accumbal shell and dorsolateral striatum which favor DA overflow in the accumbal shell. As a result, the increase in phasic burst firing, evoked by an injection of nicotine, resulted in a preferential increase in DA release in the shell, results that are consistent with the results of earlier microdialysis studies (Benwell and Balfour 1997; Cadoni and Di Chiara 2000). This group also reported that nicotine injections substantially diminish DA release evoked by low-frequency tonic activity of the neurons. Thus, phasic DA activity and release is favored following an injection of nicotine and, as a result, nicotine injections enhance the "signal-to-noise relationship" within the terminal fields. A more recent study (Li et al. 2011) suggests that an acute nicotine injection also results in the development of synchronous firing of a subset of DA neurons within the VTA, a factor that may also contribute to the substantial increase of DA overflow into the extracellular space of the accumbal shell.

The administration of cocaine is also reported to elicit a preferential enhancement of phasic burst firing of the DA neurons that project to the accumbal shell and that this contributes to the increase in DA overflow evoked by the drug (Aragona et al. 2008). An earlier study from the same group showed that both non-contingently administered cocaine and contingent self-administration of the drug enhance phasic burst firing of the mesolimbic DA neurons that project to the nucleus accumbens (Stuber et al. 2005). Non-contingent cocaine administration to both cocaine-naïve rats and rats with previous experience of the drug evoked a substantial increase in the transient release of DA which was not time-locked to specific environmental stimuli. The contingent administration of the drug also evoked the changes in phasic release which were not time-locked to a stimulus. However, a subset of the phasic release events were time-locked to the lever-pressing responses, results which are consistent with the evidence that these transients relate the lever-pressing responses to conditioned stimuli that predict drug delivery (Phillips et al. 2003). Similar studies have yet to be performed with nicotine. However, the data support the working hypothesis that the administration of psychostimulant drugs of abuse elicits characteristic changes in the firing pattern of mesoaccumbens DA neurons. Furthermore, a significant proportion of the enhanced bursting activity of the neurons does not appear to be time-locked to the presentation of relevant conditioned stimuli and may, thus, subserve a different function putatively related specifically to dependence.

4 The Role of Mesoaccumbens Dopamine Projections in Locomotor Responses to Nicotine

The DA projections from the VTA area play a complex role in the regulation of behavior. However, two aspects of behavior, stimulation of locomotor activity and responding for rewarding stimuli or outcomes, have been particularly associated with increased DA release from the mesolimbic DA neurons that project to the nucleus accumbens. The acute administration of nicotine to rats causes a dose-dependent reduction in locomotor activity during the first 20 min of the trial if the animals are placed in the test arena immediately after the injection (Clarke and Kumar 1983). However, if the animals are left in the activity cage for 80 min, the activity is increased in a dose-dependent manner during the 40–60- and 60–80-min subtrials as the animals habituate to the test environment. In a subsequent microdialysis study, Benwell and Balfour (1992) used a similar paradigm to explore the effects of acute nicotine on DA overflow in the core subdivision of the accumbens in conscious freely moving rats tested in an activity box. In this study, the animals were left for 80 min to generate stable baseline levels of extracellular DA and habituate to the box before nicotine was administered to the rats. When compared with the responses measured in control animals given the saline vehicle, an acute injection of nicotine caused a dose-dependent increase in locomotion although increased DA overflow could only be detected in these rats in the presence of an inhibitor of the DA transporter (Benwell and Balfour 1997). Pretreatment with nicotine for several days prior to the test day results in sensitization of locomotor response to the drug and evidence that the animals have become tolerant to the initial inhibitory effects of the drug on activity (Clarke and Kumar 1983). If DA overflow is measured in rats sensitized to nicotine in this way, a substantially enhanced DA response to the drug is also observed (Benwell and Balfour 1992). This observation was confirmed in a subsequent study which showed that the sensitized DA response to the drug was observed in the accumbal core but not the shell (Cadoni and Di Chiara 2000).

Other studies have shown that motor sensitization to nicotine is also observed if rats are given daily microinjections of subthreshold doses of nicotine directly into the VTA (Panagis et al. 1996). Furthermore, selective lesions of the DA projections to the nucleus accumbens are reported to attenuate or block the increases in activity observed in nicotine-tolerant rats (Clarke et al. 1988; Louis and Clarke 1998; Panagis et al. 1996). A subsequent study showed that the sensitized locomotor response to nicotine and the locomotor stimulation evoked by amphetamine are selectively attenuated by lesions of the DA projections to accumbal core but not the medial shell (Boye et al. 2001). Similar studies with amphetamine suggest that the locomotor stimulant properties of the drug depend upon the increased DA release in the accumbal core whereas the reinforcing properties of the drug, measured using a place preference paradigm, depend upon the increased DA overflow in the medial shell of the accumbens (Sellings and Clarke 2003). These observations imply that the rewarding and locomotor-activating properties of stimulants are mediated by anatomically distinct components of the accumbens and appear consistent with the hypothesis that accumbal core is implicated in the regulation of motor activity, whereas the accumbal shell forms an integral component of the limbic systems of the brain (Heimer et al. 1991).

The results summarized in the previous paragraph would appear to provide convincing support the hypothesis that the development and expression of sensitized locomotor responses to nicotine depend upon the enhanced stimulation of the DA neurons that project from the VTA and an associated increase of DA overflow in the core subdivision of the accumbens (Benwell and Balfour 1992; Cadoni and Di Chiara 2000; Heimer et al. 1991). However, a number of studies in this laboratory have cast significant doubt on this working hypothesis. The expression of the sensitized DA responses in the accumbal core is antagonized if the injections of nicotine are preceded by an injection of an NMDA receptor antagonist (Balfour et al. 1996). By contrast, pretreatment with receptor antagonists does not cause a significant attenuation of the sensitized locomotor responses to the drug. Similarly, the attenuation of the sensitized DA responses to nicotine seen in animals pretreated with ibogaine is also not associated with a reduction in the hyperlocomotion evoked by nicotine. Furthermore, pretreating animals with an NMDA receptor antagonist during the sensitization phase of the experiment also attenuates the expression of the sensitized DA response, but not the hyperlocomotion (Shoaib et al. 1994). The more recent studies have explored the effects of 6-methyl-2-(phenylethynyl)-pyridine (MPEP) on the locomotor and DA responses to nicotine. MPEP is a negative allosteric modulator at metabotropic receptor 5 (mGluR5) receptors (Rodriguez et al. 2010). These experiments suggest that the locomotor response to nicotine is also unaffected if the animals are pretreated with MPEP at a dose that attenuates the effects of nicotine on DA overflow in the accumbal shell (Tronci et al. 2010). Thus, these data suggest that the locomotor stimulation, evoked by nicotine, is not related to increased DA overflow in either subdivision of the accumbens. Other studies suggest that the sensitized motor responses to nicotine may be associated with enhanced postsynaptic responses to DA in the nucleus accumbens (Birrell and Balfour 1998; Suemaru et al. 1993). These observations have led our group to propose that the locomotor stimulant properties of nicotine may be associated with increased DA release in synapses in the nucleus accumbens, whereas the extracellular DA in this region of the brain is more concerned with other aspects of nicotine psychopharmacology, including specifically nicotine dependence (Balfour et al. 2000).

5 The Role of Mesolimbic Dopamine in Nicotine Dependence

5.1 Studies with Animal Models of Nicotine Reinforcement

There is convincing evidence that the rewarding or reinforcing properties of nicotine, measured using a self-administration paradigm, depend upon the DA release from mesoaccumbens DA neurons. This conclusion is supported by the evidence that selective lesions of the DA projections to the accumbens result in a substantial reduction of intravenous nicotine self-administration in animals that have been trained in an operant chamber to press for nicotine (Corrigall et al. 1992). Furthermore, the local administration of nicotine receptor antagonists into the VTA evokes a similar attenuation of self-administration, data that support the view that the reinforcing properties of nicotine, measured using this paradigm, depend upon the stimulation of the DA neurons that project from the VTA (Corrigall et al. 1994).

It is important to remember, however, that these observations may not be as compelling as they appear because lesioning procedures, similar to those adopted to attenuate nicotine self-administration, also attenuate the stimulatory effects of repeated nicotine on locomotor activity (Clarke et al. 1988; Louis and Clarke 1998). Moreover, the effects of intra-VTA nicotine on locomotion are inhibited by the prior administration of a nicotinic receptor antagonist (Reavill et al. 1990).

A most recent study by Cadoni et al. (2009) has shown that the self-administration of nicotine to inbred Lewis rats has a greater effect on DA overflow in both the nucleus accumbens shell and core than in the accumbens of inbred Fischer 344 rats. The study also showed that the Lewis rats were also more sensitive to the increase in locomotor activity, evoked by low doses of nicotine. A previous study suggested that, whereas Lewis rats can be trained to self-administer nicotine, Fischer 344 rats do not readily acquire this response (Brower et al. 2002). Furthermore, the Lewis rats that acquired nicotine self-administration responded preferentially on the active lever. The authors suggested that the data supported the conclusion that the Lewis strain may be genetically susceptible to the reinforcing properties of nicotine that underpin dependence. It is important to note, however, that Bower and colleagues used an extended unlimited access paradigm in which the animals were able to respond for nicotine for 23 h. By contrast, an earlier study by Shoaib et al. (1997), in which self-administration was measured over 1 h daily sessions, showed that neither Fischer nor Lewis rats readily acquired nicotine self-administration. Thus, the potential association between the genetic effects on DA overflow and reward, as measured using intravenous self-administration, should be treated with some caution.

The rewarding properties of nicotine can also be explored using a conditioned place preference paradigm in which one compartment of the apparatus is repeatedly paired with a nicotine injection. There is evidence that nicotine-induced conditioned place preference is attenuated if the DA projections to the nucleus accumbens are selectively lesioned (Di Chiara 2000). A more recent study showed that attenuation of nicotine-induced conditioned place preference could be evoked by selective lesions of the DA terminals in the accumbal shell, whereas lesions of the terminals in the accumbal core resulted in increased preference for the nicotine-paired compartment (Sellings et al. 2008). These authors concluded that the DA projections to the shell mediated the "rewarding" properties of the drug, whereas the DA projections to the accumbal core played a role in mediating the aversive properties of nicotine. However, as will be made clear later in this chapter, other roles have also been attributed to the DA neurons that project to the accumbal core.

Other researchers have employed more indirect approaches to explore the role of DA in the psychopharmacological responses to nicotine. For example, there is evidence that constituents of tobacco smoke cause inhibition of monoamine oxidase (MAO) activity in the brain and this effect is thought to enhance the addictive potential of nicotine when delivered in tobacco smoke (Brody 2006; Fowler et al. 2003). This hypothesis is supported by results reported by both Guillem et al. (2005) and Villegier et al. (2007) in which nicotine self-administration was shown to be enhanced by pretreating the animals with a drug that inhibits monoamine oxidase.

A subsequent study (Guillem et al. 2006) demonstrated that this enhancement was particularly marked in animals pretreated with a selective MAO-A inhibitor but not in animals pretreated with an MAO-B inhibitor. In contrast to human brain, in which both isoforms of the enzyme are implicated in DA metabolism, in rat brain DA is preferentially metabolized by MAO-A (Saura et al. 1992). Thus, results with experimental animals are consistent with the possibility that the enhanced motivation to respond for nicotine is associated with the potentiation of DA. In other studies, Cohen et al. (2002) showed that pretreating rats with the cannabinoid CB1 receptor antagonist, rimonabant, attenuated both the effects of nicotine on DA overflow in the nucleus accumbens shell and nicotine-seeking behavior in a self-administration study. These data provide further support for the hypothesis that increased DA overflow in the accumbal shell plays a central role in nicotine-reinforced behavior.

5.2 Studies with Other Psychostimulant Drugs of Dependence

In order to be able to interpret the results of studies with nicotine more clearly, it is instructive to consider how studies with other psychostimulant drugs of dependence have contributed to our understanding of the role of mesolimbic DA in addiction. Studies with cocaine have shown that rats can be trained to self-administer cocaine directly into the shell subdivision of the accumbens, whereas they cannot be trained to self-administer the drug directly into the accumbal core (Rodd-Henricks et al. 2002). Furthermore, this response was attenuated by the co-administration of the DA receptor antagonist, sulpiride, results which suggest that the reinforcing properties of cocaine delivered into the accumbal shell depends upon the stimulation of D2 receptors in this subdivision of the accumbens. Other studies have shown that the rewarding properties of amphetamine, measured using the place preference paradigm, are attenuated by selective lesions of the DA terminals in the accumbal shell (Sellings and Clarke 2003). Other anatomical loci have also been implicated in the reinforcing properties of psychostimulant drugs. Ikemoto (2003) noted that the DA neurons that project to the olfactory tubercle also seem to play a role in the reinforcing properties of cocaine. Furthermore, Sellings et al. (2006) extended these reported that the reinforcing properties of methylphenidate area associated with increased DA release in the medial olfactory tubercle. Thus, there is a need to be cautious when attributing the locomotor stimulant and reinforcing properties of nicotine solely to the events in the two principal subdivisions of the nucleus accumbens.

While the evidence that DA neurons play a central role in the reinforcing properties of psychostimulant drugs that underpin dependence is reasonably compelling, some studies suggest that the relationship is not simple. For example, Rocha et al. (1998) showed that transgenic mice lacking function neuronal DA transporters were hyperactive and could still learn to self-administer cocaine in manner that seemed similar to that seen for wild-type animals. The authors concluded that the reinforcing properties of cocaine cannot depend solely on its ability to inhibit the DA transporter and that other neural responses also play a key role in cocaine reinforcement. Studies with other drugs of abuse also suggest that DA may play a specific role in drug reinforcement. There is evidence that mesolimbic DA neurons may facilitate acquisition of opiate self-administration in rodents (Bozarth and Wise 1984; Singer and Wallace 1984). However, Pettit et al. (1984) demonstrated that, once the response had been acquired, lesions of the DA projections to the accumbens have no significant effects of heroin self-administration, whereas the lesions did attenuate established cocaine self-administration. Similar findings were reported by Gerrits and Van Ree (1996) who suggested that the results argued against a critical role for mesolimbic DA in the motivational mechanisms underpinning drug dependence. A subsequent study by Cannon and Palmiter (2003) showed that transgenic mice lacking tyrosine hydroxylase, and therefore the ability to synthesize DA, could still learn to respond for a sucrose reward. Thus, the data taken together support the possibility that increased DA release in the nucleus accumbens may facilitate the acquisition of responding for rewards or reinforcers but that it is not an absolute requirement.

5.3 Does Extracellular Dopamine in the Accumbal Shell Have a Specific Role in Nicotine Dependence?

Since a majority of drugs of dependence stimulate DA release and/or sustain DA overflow into the extracellular space, it is long been argued that this property of the drugs contributes significantly to neuropharmacological mechanisms which underpin the development of dependence (Di Chiara 1995; Wise 1987). Drug dependence is a chronic relapsing condition and it follows that the changes in the activity of mesolimbic DA neurons which occur as a consequence of chronic exposure to these drugs is likely to contribute to the mechanisms that underpin the transition to dependence. As already discussed, daily injections of nicotine result in sensitization of DA overflow in the core subdivision of the accumbens (Benwell and Balfour 1992; Cadoni and Di Chiara 2000; Iyaniwura et al. 2001). By contrast, the repetitive administration of nicotine, using the same protocol, does not elicit sensitization of DA overflow in the accumbal shell or dorsolateral striatum (Benwell and Balfour 1997; Cadoni and Di Chiara 2000; Iyaniwura et al. 2001). Indeed, some studies have shown that the effects of nicotine on DA overflow in the shell subdivision exhibit tolerance to nicotine if the drug is administered repeatedly (Cadoni and Di Chiara 2000; Tronci and Balfour 2011). If the animals are allowed to self-administer nicotine, a different pattern of responses is observed in the two principal subdivisions of the accumbens (Lecca et al. 2006). During the first week of nicotine self-administration, the effects of the drug on DA overflow in the nucleus accumbens shell are clear and significant. By contrast, the effects on DA overflow in the accumbens core are modest and do not achieve statistical significance. Thus, the effects of self-administered nicotine on DA overflow in the principal subdivisions of the accumbens resemble those seen in animals treated acutely with nicotine. During the third week of nicotine self-administration, there is a modest but statistically significant sensitization of the DA overflow in the accumbal core. However, in contrast to the effects of non-contingent nicotine, three weeks of self-administered nicotine result in sensitization of its effects on DA overflow in the accumbal shell. Similar sensitization of DA overflow in the accumbal shell is also observed in animals trained to self-administer cocaine or heroin (Lecca et al. 2007a, b). These observations support the view that preferential sensitization of the increase in DA overflow in the accumbal shell is a specific characteristic associated with the acquisition of contingent drug self-administration and play an important role in behaviors associated with the drug-taking behavior.

The putative behavioral significance of the changes in extracellular DA remains to be established. Grace (2000) has argued that the primary role of extracellular DA is to regulate phasic DA release through the DA autoreceptors and that the persistent increase in extracellular DA, evoked by chronic exposure to drugs of abuse, results in a sustained suppression of phasic responses. He posits that the craving to take drugs of abuse reflects a need to overcome this increased inhibitory effect and, thus, restore normal phasic DA release. Others have argued that increased extracellular DA, especially in the accumbal shell, plays a more direct role in drug-seeking behavior (Balfour 2009; Di Chiara et al. 2004). For nicotine, this hypothesis receives support from a number of studies which report that treatments which inhibit the increase in extracellular DA in the nucleus accumbens shell, evoked by the drug, also attenuate nicotine self-administration and conditioned place preference (Balfour 2009; Cohen et al. 2005, 2002; Di Chiara et al. 2004; Scherma et al. 2012, 2008; Tronci and Balfour 2011). Furthermore, in animals in which responding for the drug has been extinguished, the reinstatement of nicotine-seeking behavior, evoked by a non-contingent injection of nicotine, is also attenuated by a treatments that attenuate the increase in DA overflow in the accumbal shell evoked by nicotine (Mascia et al. 2011; Scherma et al. 2008). None of the studies have sought to determine whether the treatments act selectively or preferentially on phasic or tonic firing of the neurons.

The results presented above are consistent with the hypothesis that, like other psychostimulant drugs of abuse, increased DA overflow in the accumbal shell mediates the "rewarding" properties of nicotine. However, Salamone and colleagues (Salamone and Correa 2013; Salamone et al. 2012) have summarized a considerable body of evidence which suggests that responding for natural rewards, such as food, does not depend upon the stimulation of the DA projections to the nucleus accumbens. These pathways, however, do enhance behavioral activation, the motivational salience of the rewards, and effort-related choice. This hypothesis is consistent with an earlier study which demonstrated that transgenic mice, lacking a key enzyme (tyrosine hydroxylase) in the biosynthetic pathway for DA, still exhibit robust preference for sucrose over water although the transgenic animals consumed less sweetened solution than the wild-type controls (Cannon and Palmiter 2003). The authors concluded that the mice lacking DA exhibited a deficiency in goal-directed behavior rather than an impaired perception of the reward. Thus, there is a growing consensus that the DA projections to the accumbens influence the motivation to respond for a reinforcer, especially in an instrumental paradigm, rather than mediating their rewarding properties *per se*. It also seems reasonable to suggest, therefore, that compulsive drug-seeking and drug-taking behaviors reflect the effects of addictive drugs on these processes (Balfour 2009; Salamone and Correa 2012; Salamone et al. 2007). Salamone (Nunes et al. 2013; Salamone et al. 2012) and his colleagues have focused on the possibility that increased DA overflow into the extracellular space of the nucleus accumbens, including that evoked by drugs of dependence, enhances behavioral activation and effort-related choice behavior for rewards. By contrast, psychopathological conditions, such as depression, are posited to be associated with reduced DA release in this area of the brain and that this effect mediates the psychomotor slowing, anergia and apathy that characterize these conditions.

The increases in extracellular DA in the nucleus accumbens, evoked by the administration of nicotine and other drugs of dependence, are generally sustained for periods up to an hour or more. Thus, it seems reasonable to posit that they mediate the effects of sustained diffuse stimuli on drug-seeking or drug-taking behavior that are not time-locked to specific-conditioned stimuli (Balfour 2009). Studies in a number of laboratories have shown that nicotine self-administration is attenuated by pretreating the animals with an mGluR5 receptor antagonist (Palmatier et al. 2008; Paterson et al. 2003; Tessari et al. 2004). Tronci and colleagues (Tronci and Balfour 2011; Tronci et al. 2010) showed that this attenuation of nicotine self-administration is evoked at doses of antagonist that also attenuate the increase in DA overflow evoked by nicotine in the accumbens shell (Fig. 2). These data would appear to support the hypothesis that increased DA overflow in the accumbal shell mediates nicotine reinforcement in this instrumental paradigm and that the response to nicotine depends upon the co-stimulation of mGluR5 receptors. A parallel study by D'Souza and Markou (D'Souza and Markou 2011) showed that the microinfusion of an mGluR5 receptor antagonist directly into the VTA or accumbal shell attenuated nicotine self-administration. These results provide further support for a role for the DA projections to the accumbal shell in the reinforcing effects of nicotine and implicate mGluR5 receptors in these areas of the mesolimbic system in the response. However, the microinfusions into the VTA, but not the accumbal shell, also attenuated responding for a food reward. Systemic injections of moderate doses of MPEP, which attenuate nicotine self-administration, do not inhibit responding for a palatable food reward (Bespalov et al. 2005; Tronci and Balfour 2011) although it does attenuate the motivation to respond for food as determined using a progressive ratio paradigm (Paterson and Markou 2005). The effects of MPEP microinjections into the accumbal shell, therefore, most closely mimic the effects of systemic drug on nicotine self-administration, and it seems reasonable to conclude that the accumbal shell may be the primary site of the antagonists when they are administered systemically.

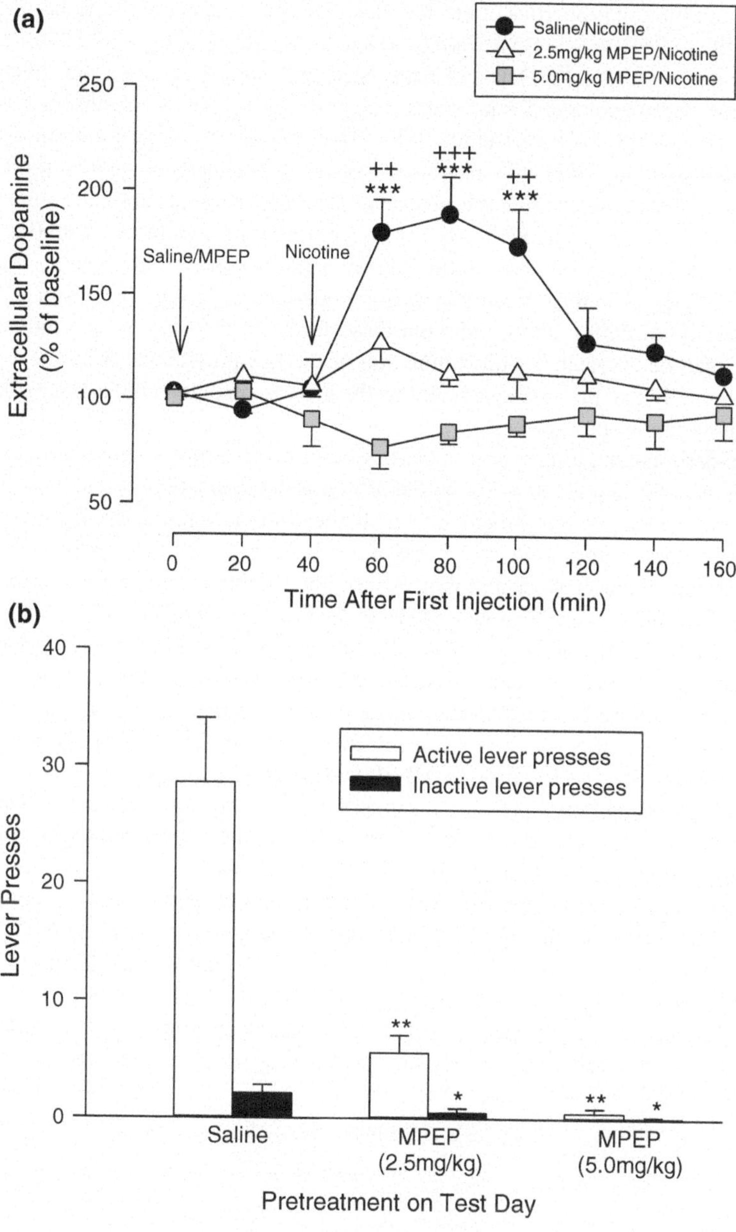

Fig. 2 The effects of MPEP on dopamine overflow in the nucleus accumbens and nicotine IVSA. Panel **a** show the effects of pre-injecting animals with MPEP (2.5 or 5.0 mg/kg IP) or its saline vehicle on the increase in DA overflow in the shell subdivision of the nucleus accumbens evoked by an acute SC injection of nicotine (0.4 mg/kg). Panel **b** shows the effects of the same doses of MPEP on lever-pressing in rats trained to respond for IV nicotine (30 μg/kg). The results are presented as means ± sem. Significantly different from rats pretreated with saline (**$p < 0.01$; ***$p < 0.001$). The figure is adapted from Tronci and Balfour (2011)

The reinforcing properties of nicotine that underpin dependence have been related to its ability to enhance brain reward function, as measured using an intracranial self-stimulation paradigm (O'Dell and Khroyan 2009) and to enhance the salience of other non-pharmacological reinforcers (Caggiula et al. 2002; Chaudhri et al. 2006; Donny et al. 2003; Palmatier et al. 2006). The reward-enhancing properties of nicotine are also observed in animals trained to self-administer the drug directly into the VTA, data which suggest that this property of the drug depends upon the stimulation of neurons, putatively DA neurons, that project from this area of the brain (Farquhar et al. 2012). Pretreatment with an mGluR5 receptor antagonist, however, is reported to have no effects on the reward-enhancing properties of nicotine (Kenny et al. 2003; Palmatier et al. 2008). Thus, although these facets of the nicotine psychopharmacology may be associated with stimulation of the DA projections from the VTA, it seems unlikely that they are dependent on the increase in extracellular DA evoked by the drug in the accumbal shell.

In experimental animals, both sensitization of the locomotor stimulant properties of nicotine and nicotine-seeking behavior in a self-administration paradigm can be conditioned to the environment in which the drug is administered (Bevins and Palmatier 2003; Diergaarde et al. 2008; Reid et al. 1998; Wing and Shoaib 2008). Furthermore, there is evidence that distal cues of this nature exert a significant influence on the craving to smoke tobacco (Le Foll and Goldberg 2005; Van Gucht et al. 2010). In the nicotine self-administration studies reported by Tronci and Balfour (2011), pretreatment with the mGluR5 negative allosteric modulator, MPEP, also suppressed responding on the inactive lever (Fig. 2). This observation implies that the drug does not act selectively on the reinforcing properties of nicotine. Moreover, as the dose of MPEP was increased, a higher proportion of the rats tested made no lever-pressing responses at all during the 1-h session (Table 1). This aspect of the response to MPEP, therefore, cannot be attributed to a neural interaction between nicotine and the mGluR5 antagonist. Moreover, studies on the effects of MPEP on locomotor activity indicated that the deficits in responding on the instrumental task did not reflect a general reduction in activity. MPEP pretreatment, however, selectively attenuated contextually conditioned hyperactivity in rats repetitively exposed to a maze following each daily injection of nicotine while having no significant effects on context-independent pharmacological sensitization of the locomotor response to the drug (Fig. 3) (Tronci et al. 2010).

Table 1 The influence of MPEP on nicotine intravenous self-administration

MPEP dose (mg/kg)	Percentage of rats responding for nicotine
Saline	100
2.5	88
5.0	38

The table depicts the percentage of rats that made at least one lever-pressing response during a 1-h session following the administration of the saline vehicle or increasing doses of MPEP. Each rat ($n = 8$) was tested in the operant chamber following an IP injection of MPEP or its saline vehicle using a counter-balanced design. The injections were given 30 min prior to testing in the operant chamber

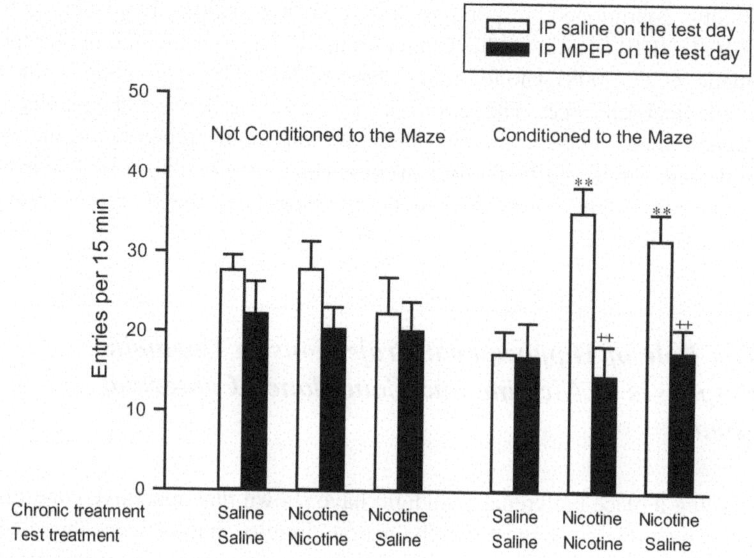

Fig. 3 The influence of MPEP on contextually conditioned locomotor sensitization to nicotine. Groups of rats was given daily injections of saline or nicotine (0.4 mg/kg SC) for 16 days. Half the rats in each treatment group were returned to their home cages; the remainder were place in a 4-arm maze for 15 min. On days 17 and 21, the rats were given IP saline (*open columns*) or MPEP (5 mg/kg; *filled columns*) using a counter-balanced design. The chronic treatment and the treatment on the test day are shown along the *x*-axis of the graphs. The figure shows the activity of the rats (entries into the arms of the maze) expressed as mean ±sem of 6 observations for rats that were habituated to saline or nicotine in the home cage or habituated to the maze after each injection. ** significantly different from saline/saline group: $p < 0.01$; ++ significantly different given IP saline: $p < 0.01$. Taken from Tronci et al. (2010) with permission

These observations seem most consistent with the possibility that a principal effect of MPEP is the attenuation of contextually conditioned behavioral responses to nicotine. Thus, MPEP seems to inhibit behavioral responses to nicotine in two ways. It attenuates the stimulation of DA overflow evoked in the nucleus accumbens by nicotine and, thereby, behaviors such as self-administration that depend upon this neural response. It also attenuates the effects of distal contextual cues that act, putatively through the hippocampus, to modulate the effects of nicotine on mesolimbic function and to promote nicotine-seeking behavior.

The effects of MPEP on the reward-enhancing properties of nicotine have also been explored using a paradigm similar to that described by Palmatier et al. (2007). MPEP pretreatment attenuated the enhanced responding for a non-pharmacological complex visual reinforcer seen in the animals given nicotine but also attenuated the moderate level of responding for the non-pharmacological reinforcer seen in animals trained with saline. As with the nicotine self-administration study, the effects of MPEP were not selective for the active lever. The data generated by this study, therefore, would appear consistent with the hypothesis that MPEP also attenuates contextually conditioned responding for this non-pharmacological reinforcer (Tronci et al. 2010). However, the data presented above contrast with those reported by Palmatier et al. (2008) who found no effects of MPEP on responding for the non-pharmacological reinforcer. The reason for the difference between the results of the two groups remains unclear although there are important differences in the experimental designs employed by the two groups which may explain why contextually conditioned responding contributed more significantly to the study performed by Tronci and Balfour (2011).

5.4 The Role of Hippocampal Projections in Dopamine Responses to Nicotine and Conditioned Contextual Stimuli

Studies using a place preference paradigm have shown that appetitive conditioning using a sucrose reward suggest that the neural circuitry that mediates the effects of conditioned spatial contextual cues on reward-seeking behavior differ from the effects of discrete conditioned stimuli (Ito et al. 2008). Significantly, these studies revealed that excitotoxic lesions of the accumbal shell attenuated the effects of the contextual cues, whereas lesions of accumbal core attenuated responding for discrete conditioned stimuli paired with presentation of the reward. The role of the accumbal shell depended upon the intact neural connections to the hippocampus. These results support the conclusion that limbic corticostriatal network plays a central role in spatial contextual conditioning. Moreover, studies with drugs of dependence suggest that the effects of spatial contextual stimuli on drug-seeking behavior are associated particularly with increased DA release in the accumbal shell (Bossert et al. 2007; Chaudhri et al. 2010; Crombag et al. 2008; Ito et al. 2000). DA overflow in the nucleus accumbens is increased by stimulation of glutamatergic projections from the hippocampus (Floresco 2007; Floresco et al. 2001; Legault and Wise 1999; Taepavarapruk et al. 2000). It has been proposed that these glutamatergic pathways play a pivotal role in the effects of distal contextual cues on behavior (Crombag et al. 2008). Crombag and colleagues focused on the role of glutamatergic projections from the dorsal hippocampus to the nucleus accumbens, whereas Grace and colleagues (Grace et al. 2007) have emphasized the role of the projections from the ventral hippocampus. The latter system is reported to enhance DA overflow in the nucleus accumbens shell via a circuit which projects through the ventral pallidum to the VTA by increasing the proportion of DA neurons that are tonically active. These observations imply that distal contextual cues may enhance the release of DA into the extracellular space of the shell of the accumbens by increasing the tonic activity of DA the neurons in the VTA. Stimulation of the glutamatergic afferents from the ventral subiculum to the nucleus accumbens increases both DA overflow and locomotor activity (Taepavarapruk et al. 2000). Studies to date have implicated both NMDA and non-NMDA glutamatergic receptors in this circuit (Floresco 2007; Floresco et al. 2001; Legault and Wise 1999; Taepavarapruk et al. 2000). There is a paucity of studies that have sought to explore the putative role of the hippocampus and its connections to the nucleus accumbens on nicotine-evoked contextual conditioning and, currently, there is no direct evidence that mGluR5 receptors are also implicated in the role of hippocampal afferents on DA release in the nucleus accumbens. However, the accumbal neurons that innervate the ventral pallidum are rich in mGluR5 receptor RNA, whereas only 50 % of the neurons that project directly to the VTA express these receptors (Lu et al. 1999), a finding that provides some indirect evidence for a role for mGluR5 receptor involvement in the circuit.

5.5 *The Role of the Extracellular Dopamine in the Accumbal Core*

The repetitive non-contingent administration of drugs of nicotine results in a subregionally selective sensitization of its effects on DA overflow into the extracellular space of the accumbal core, an effect that nicotine shares with other drugs of dependence (Cadoni and Di Chiara 1999, 2000; Cadoni et al. 2000; Iyaniwura et al. 2001). Both the acquisition and expression of this sensitized DA response are attenuated by pretreatment with drugs that antagonize NMDA glutamatergic receptors, data which suggest that the sensitized response depends upon the co-stimulation of these receptors (Balfour et al. 1996; Shoaib et al. 1994). Other data support the conclusion that the sensitized DA response is caused by increased phasic burst firing the neurons that project to the accumbal core (Balfour et al. 1998; Schilstrom et al. 1998). This hypothesis is supported by a more recent study which demonstrated that increased phasic burst firing of the DA projections to the accumbal core depends upon the co-stimulation of nicotinic receptors composed of $\alpha6\beta2$ subunits and NMDA receptors located on the DA neurons (Wickham et al. 2013). Data presented earlier in the chapter excluded the possibility that the increase in DA overflow, evoked by nicotine in the accumbal core, mediates the sensitized locomotor response seen in these animals when they are challenged with nicotine. These findings pointed to the likelihood that the sensitized DA responses mediate some other aspects of nicotine psychopharmacology, putatively associated with dependence (Balfour 2009; Balfour et al. 2000) although the nature of this role remains a matter of speculation (Di Chiara 2002; Di Chiara and Bassareo 2007). The administration of an NMDA receptor antagonist has, however, been shown to attenuate both nicotine self-administration and the increased brain reward function elicited by a nicotine injection (Kenny et al. 2009). Reduced nicotine self-administration and nicotine-facilitated brain reward function has also been observed in animals in which the antagonist was delivered locally into the VTA (Kenny et al. 2009). Thus, the reinforcing and reward-enhancing properties of nicotine can speculatively be associated with increased burst firing of mesolimbic DA neurons, but not necessarily those which project to the accumbal core. Indeed a subsequent study (D'Souza and Markou 2014) has shown that the responding for a nicotine-associated conditioned cue is enhanced by the local administration of an NMDA receptor antagonist into the accumbal core.

Discrete conditioned stimuli, repetitively paired with the self-administration of cocaine, substantially enhance self-administration of the drug and selectively activate neurons in the accumbal core (Hollander and Carelli 2007; Ranaldi and Roberts 1996). Selective excitotoxin-evoked lesions of the accumbal core have little effect on schedule-controlled cocaine self-administration but substantially inhibit drug-seeking behavior maintained by drug-associated conditioned cues (Ito et al. 2004). Other studies, reported by the same group, have shown that cocaine-seeking behavior, maintained by cocaine-paired conditioned stimuli, has no effects on DA overflow in the accumbal core or shell, but that non-contingent presentation of these stimuli results in a regionally selective increase in DA overflow in the core (Ito et al. 2000). In contrast, operant responding for cocaine-associated conditioned stimulus is associated with increased extracellular DA in the dorsal striatum, whereas the non-contingent presentation of the cue had no significant effects on extracellular DA in this region of the brain (Ito et al. 2002). These observations, the authors argued, suggest that increased extracellular DA in the accumbal core might be implicated in the reinstatement of drug-seeking behavior, evoked by exposure to drug-associated cues, whereas increased DA overflow in the dorsal striatum may be implicated in the maintenance of habitual cocaine-seeking behavior. These conclusions are consistent with recent results, employing an optogenetic approach, which show that cue-evoked reinstatement of cocaine-seeking behavior seems to be associated specifically with stimulation of the indirect pathway from the accumbal core to VTA that passages through the dorsoventral pallidum (Stefanik et al. 2013).

The self-administration of nicotine in experimental animals is also enhanced significantly if delivery of the drug is paired with a conditioned sensory stimulus (Caggiula et al. 2002, 2001). Few studies, however, have explored the specific role of the accumbal core in this aspect of nicotine reinforcement, and none has focused on the role of DA release specifically. Bassareo et al. (2007) have reported that a conditioned stimulus associated with the non-contingent presentation of nicotine preferentially stimulates DA overflow in the accumbal shell rather than the core. It is important to note, however, that this study did not involve an instrumental response or seek to measure nicotine-seeking behavior. Thus, its putative relevance to the role of conditioned stimuli in nicotine self-administration remains speculative.

6 Evidence for Tolerance to the Effects of Nicotine on Dopamine Release in the Nucleus Accumbens

In the studies discussed so far, nicotine was either administered non-contingently as a series of daily injections or contingently in daily nicotine self-administration sessions. Nicotine is rapidly metabolized by the rat and, therefore, these protocols result in exposure to nicotine which is of fairly limited duration (Kyerematen et al. 1988).

Moreover, the blood nicotine concentrations will be vanishingly low on each occasion nicotine is initially administered at the beginning of each day. However, many habitual smokers maintain raised plasma nicotine concentrations for a significant part of the day (Benowitz et al. 1990). Thus, other chronic treatment protocols have sought to model this pattern of exposure to nicotine more closely by administering nicotine over a longer period each day. One of the most common approaches adopted employs subcutaneous osmotic minipumps to deliver nicotine constantly for periods up to four weeks. It has been argued that this protocol renders the animals dependent upon the nicotine because the abrupt withdrawal of the drug elicits behavioral changes which are thought to model components of the abstinence syndrome experienced by habitual smokers when they first quit smoking (Kenny and Markou 2001; Malin 2001; Malin and Goyarzu 2009; Malin et al. 1992) (see chapter entitled Nicotine Withdrawal; this volume). Moreover, the abrupt withdrawal of nicotine in this model also decreases brain reward function, as assessed using intracranial self-stimulation (Epping-Jordan et al. 1998). This deficit is thought to be a physiological correlate of the anhedonia experienced by many abstinent smokers (D'Souza and Markou 2010; Paterson and Markou 2007). The behavioral measures of withdrawal can also be evoked by the administration of nicotinic receptor antagonists to animals constantly infused with nicotine (Epping-Jordan et al. 1998; Hildebrand et al. 1997; Malin et al. 1994). The changes in reward threshold, evoked by nicotine withdrawal, are mediated centrally, whereas the somatic signs of withdrawal involve both central and peripheral mechanisms (Hildebrand et al. 1997; Watkins et al. 2000).

A majority of the studies that have investigated the effects of the constant infusion of nicotine from osmotic minipumps on extracellular DA in either the core or shell subdivisions of the accumbens have found no effects (Balfour et al. 2000; Benwell et al. 1995; Hildebrand et al. 1998; Paterson et al. 2007) although one study did report a significant increase in basal extracellular DA in the accumbal shell (Carboni et al. 2000). The constant infusion of nicotine from minipumps, at doses that generate plasma concentrations of nicotine that are relevant to tobacco smoking, attenuates or blocks the increases in extracellular DA evoked by an injection of nicotine (Balfour et al. 2000; Benwell et al. 1995; Carboni et al. 2000). The infusions also attenuate behavioral responses to nicotine, locomotor stimulation, and intravenous self-administration, which depend upon the stimulation of mesolimbic DA neurons (Benwell et al. 1995; Coen et al. 2009). Moreover, if an acute injection of nicotine is followed by another injection within 60–90 min of the first injection, the influence of the second injection on DA overflow in the accumbal shell is substantially blunted (Balfour et al. 2000). These data support the conclusion that raised plasma nicotine levels result in tolerance to a subsequent challenge with the drug.

The withdrawal of nicotine, evoked in rats by removing the pumps, also has no significant effects on extracellular DA in either the accumbal core or shell (Benwell et al. 1995; Paterson et al. 2007). Thus, the behavioral responses evoked by removal of the pumps, following a period of constant infusion, do not seem to be associated with a profound reduction in extracellular DA concentration in the accumbens.

This is a surprising finding because Grieder et al. (2012) reported that mice, constantly infused with nicotine from an osmotic minipump, showed decreased tonic but not phasic firing of mesoaccumbens DA neurons and that this was enhanced when nicotine was withdrawn. This contrasts with the effects of acute nicotine administration to drug-naïve animals that selectively stimulates the number and length of phasic burst firing episodes (Zhang et al. 2009). This study also found that a behavioral measure of nicotine withdrawal (conditioned place aversion) was related to reduced stimulation of the DA D2 receptors that mediate the responses to tonic DA release. Zhang and colleagues (Zhang et al. 2012) reported that the withdrawal of nicotine from mice, treated chronically with the drug in their drinking water, elicits a reduction in DA release mediated by tonic firing of the neurons. In this study, however, withdrawal did not exert a selective effect on tonic DA release although its effects on phasic DA release were less marked. In contrast to the results with minipumps, in this study nicotine withdrawal also reduced extracellular DA in the accumbens shell.

The attenuation of the behavioral and DA responses to a nicotine challenge are consistent with the possibility that sustained exposure to the drug results in desensitization of the nicotinic receptors that mediate its effects on mesoaccumbens DA neurons (Balfour et al. 1998; Benwell et al. 1995). This hypothesis is supported by the results of electrophysiological studies which showed that exposing DA neurons in the VTA to nicotine results in transient stimulation and then desensitization of the neuronal nicotinic receptors located on these neurons (Pidoplichko et al. 1997). However, the systemic administration of a nicotinic receptor antagonist to rats that have been constantly infused with nicotine from an osmotic minipump reliably depresses the basal levels of DA in the extracellular space of the accumbal shell (Carboni et al. 2000; Hildebrand et al. 1998; Pidoplichko et al. 1997; Rada et al. 2001). Similar effects on behavior and the extracellular concentration of DA in the accumbal shell are also evoked by microinjecting mecamylamine directly into the VTA (Hildebrand et al. 1998), whereas microinjections of the antagonist into the nucleus accumbens do not elicit this effect (Hildebrand and Svensson 2000). Similarly, the microinjection of the more selective neuronal nicotinic receptor antagonist, dihydro-β-erythroidine, directly into the VTA but not the accumbens elicits the deficits in brain reward that are also observed following the removal of the pumps (Bruijnzeel and Markou 2004). These results suggest that the behavioral and neural tolerance to a nicotine challenge, observed in animals constantly infused with nicotine, is not solely associated with receptor desensitization but is also likely to reflect a neuroadaptive increase in the inhibitory control of mesolimbic DA neurons at the level of the VTA. This possibility is consistent with the evidence, cited above, that the constant infusion of nicotine results in inhibition to tonic activity in mesolimbic DA neurons which is enhanced when the drug is withdrawn (Grieder et al. 2012; Zhang et al. 2012). Reduced DA overflow in the nucleus accumbens is not universally observed following the withdrawal of other drugs of dependence and unlikely to be a common feature of drug withdrawal (Crippens et al. 1993; Crippens and Robinson 1994; Grieder et al. 2012; Zhang et al. 2012). However, the effects of chronic exposure to psychostimulant drugs on the regulation of mesolimbic DA neurons seems to depend upon the chronic regimen employed. Like nicotine, the abrupt withdrawal of amphetamine or cocaine, following a period of continuous exposure, results in increased brain reward threshold when measured using intracranial self-stimulation (Cryan et al. 2003; Markou and Koob 1991). The effects of abrupt drug withdrawal on extracellular DA in the nucleus accumbens, when the drugs are administered using these treatment regimes, have not been explored.

In nicotine-withdrawn mice, the acute administration of nicotine by means of an injection resulted in the extracellular DA levels in the accumbal shell temporarily returning to control levels (Zhang et al. 2012). Since the chronic treatment with nicotine depressed basal DA levels in these animals, when expressed as a percentage of the baseline value, the responses to the nicotine injection were enhanced in the withdrawn rats. Benwell and colleagues (Balfour et al. 1996) also found evidence for sensitization of the DA response to nicotine in the accumbal core that was statistically significant when measured 2 or 7 days after removal of the pumps. Moreover, the locomotor response to the nicotine challenge was also significantly enhanced by nicotine withdrawal, this sensitization achieving statistical significance within 24 h of removing the pumps.

7 The Implications for Our Understanding of the Psychobiology of Tobacco Dependence

The data summarized in this chapter support the conclusion that the DA projections to the nucleus accumbens play a central role in mediating the reinforcing properties of nicotine that underpin intravenous self-administration of the drug in experimental animals. Studies employing positron emission tomography to measure the displacement of the DA D1 receptor antagonist [11C]-SCH23390 and the D2 receptor antagonist [11C]-raclopride have shown that inhaling tobacco smoke also stimulates DA release in the ventral striatum of human brain (Brody et al. 2004; Dagher et al. 2001; also see chapter entitled Imaging Tobacco Smoking with PET and SPECT; this volume). A subsequent study showed that the effects of smoking on DA release, measured using this procedure, are dependent upon the "dose" of tobacco inhaled (Brody et al. 2010). Thus, inhaling nicotine in the form of tobacco smoke elicits changes in DA release that resemble those seen in rats given nicotine injections. The putative significance of these findings to the reinforcing properties of tobacco smoke is enhanced by the report that pretreating smokers with the D2 receptor antagonist, haloperidol, increased the amount of nicotine inhaled as the smokers, it was postulated, compensated for the diminished rewarding effects of the smoke evoked by antagonizing the receptors (Dawe et al. 1995).

However, it seems unlikely the increases in extracellular DA, evoked by nicotine, provide a complete explanation for the sustained self-administration of nicotine seen in experimental animals or human smokers because self-administration of the drug rapidly leads to the accumulation of blood nicotine levels similar to those required to attenuate the effects of a nicotine injection on DA overflow in the accumbens (Balfour et al. 2000; Benwell et al. 1995; Shoaib and Stolerman 1999). Furthermore, the constant infusion of nicotine from an osmotic minipump, at doses similar to those employed by Balfour and colleagues, attenuates responding for nicotine in a self-administration experiment (Coen et al. 2009; LeSage et al. 2003; Paterson and Markou 2004). The data imply that any effects of inhaled nicotine in tobacco smoke on DA release in the nucleus accumbens are likely to be substantially blunted as the plasma concentration of the drug increases during the day and that other non-DAergic mechanisms reinforce tobacco smoking at these times (Balfour 2009).

In experimental animals, conditioned stimuli closely paired with the delivery of nicotine also significantly enhance nicotine self-administration and can maintain nicotine-seeking behavior in the absence of the drug (Caggiula et al. 2001). Moreover, these stimuli reinstate nicotine-seeking behavior more effectively than priming doses of nicotine following extinction (Caggiula et al. 2001; LeSage et al. 2004). Caggiula and colleagues suggested that these stimuli may be at least as important as nicotine in maintaining smoking behavior. Rose and colleagues (2000) showed that habitual smokers given a de-nicotinized cigarette to smoke reported that this cigarette significantly reduced craving and was rated significantly more satisfying and rewarding than the no-smoking condition. These and similar results led Rose (2006) to hypothesize that, while nicotine plays an essential role in tobacco dependence, the sensory properties of tobacco smoke are also pivotally important in tobacco dependence. The stimuli seem to be particularly important for highly dependent smokers (Brauer et al. 2001). These observations led Balfour (2009) to suggest that tobacco smoking might best be thought of as a second schedule of reinforcement in which the primary reinforcing properties of nicotine are only experienced when the plasma nicotine concentration is low enough to permit stimulation of DA release in the accumbens (Fig. 4). It is posited that the increased DA release, evoked by nicotine at these times, is also required to maintain the salience of conditioned sensory stimuli associated with the delivery of nicotine. Once the plasma nicotine concentration is raised, the sensory stimulation when tobacco smoke is inhaled becomes the principal rewarding factor that reinforces the habit although the attenuation of withdrawal may also play a part. If this hypothesis is correct, it suggests that smoking throughout the day may represent the optimum way for an addicted individual to use tobacco (Balfour 2009). Controversially, the hypothesis also predicts that smokers may continue to find inhaling tobacco smoke "rewarding" when it is inhaled by people using nicotine replacement therapies, such as the nicotine patch or gum, that elicit sustained increases in blood nicotine. This could explain why these forms of nicotine replacement therapy tend to be the least efficacious (Stead et al. 2012) and why a proportion of smokers using these pharmacotherapies continue to smoke, albeit at a reduced level (Beard et al. 2013; Stead and Lancaster 2007).

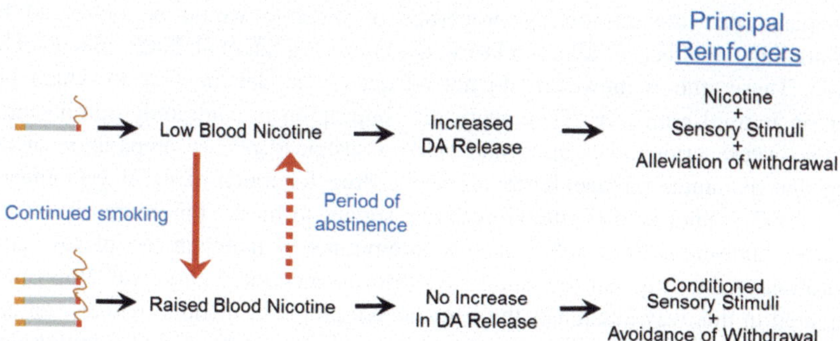

Fig. 4 A proposed second-order schedule of reinforcement for tobacco smoking. This figure depicts the nature of the reinforcers that rewards smokers when they smoke a cigarette when the plasma nicotine concentration is low after a period of abstinence (e.g., over night) and during periods when the plasma nicotine concentration is raised by prior exposure to tobacco smoke inhaled earlier in the day

7.1 *The Role of Mesolimbic Dopamine in Pharmacotherapy for Tobacco Dependence*

A majority of the pharmacological approaches employed to treat tobacco dependence focus primarily on the treatment of nicotine withdrawal and, thereby, facilitate successful cessation attempts. This is clearly true for nicotine replacement therapies. In animal studies, injections of nicotine elicit a temporary attenuation of the somatic symptoms of withdrawal and its effects on brain reward thresholds (Epping-Jordan et al. 1998; Kenny and Markou 2001; Malin and Goyarzu 2009; Malin et al. 1992). Nicotine injections also temporarily reverse the reduction in extracellular DA observed in the accumbens of mice following the withdrawal of chronic nicotine (Zhang et al. 2012). Thus, the results of animal studies support the conclusion that the both the peripheral and central effects of nicotine withdrawal can be ameliorated by the administration of nicotine. It is important to note, however, that the animal experiments do not model well the potential effects of nicotine replacement therapies, such as the nicotine patch, that deliver drug slowly. Interestingly these preparations tend to be less efficacious than preparations, such as nicotine nasal or oral sprays, that deliver nicotine to the blood stream more quickly and may be more similar in their pharmacokinetic profile to a nicotine injection (Kraiczi et al. 2011; Stead et al. 2012; Sutherland et al. 1992).

The other two principal pharmacotherapies for tobacco dependence are bupropion and varenicline. Bupropion inhibits the neuronal transporters for DA and noradrenaline and antagonizes the effects of nicotine at nicotinic receptors. It seems likely that these effects represent the principal mechanisms underpinning the efficacy of the drug as an aid to smoking cessation (Warner and Shoaib 2005). The administration of low or moderate doses of bupropion enhances nicotine self-administration in rats. This observation is not unexpected because it is to be anticipated that the reinforcing properties of nicotine would be enhanced by potentiating the effects of released DA in the brain (Rauhut et al. 2003; Shoaib et al. 2003). These studies, however, do not adequately model the way in which bupropion is used clinically. Therapeutically, bupropion is commonly administered for two weeks prior to cessation and using a sustained release preparation of the drug that maintains therapeutic levels of the drug for much of the day (Johnston et al. 2002). Other studies, therefore, have sought to model this more closely by infusing bupropion from subcutaneous minipumps to maintain the plasma concentration of the drug starting some days prior to nicotine withdrawal. Bupropion delivered in this way attenuates the somatic symptoms of nicotine withdrawal and attenuates the deficit in brain stimulation reward evoked by nicotine withdrawal (Malin et al. 2006; Paterson et al. 2007). Microdialysis studies suggest that this response to bupropion is not associated with an increase in extracellular DA in the accumbal shell (Paterson et al. 2007). However, the increase in DA overflow, evoked by the application of a depolarizing concentration of K^+, was enhanced in the animals treated chronically with bupropion and the drug also attenuated the modest reduction in evoked DA overflow observed in nicotine-withdrawn rats. The results, the authors argued, support the hypothesis that the efficacy of bupropion as an aid to smoking cessation may lie principally with its ability to attenuate the effects of nicotine withdrawal and depend upon its ability to inhibit the reuptake of DA in the brain.

The newest and, perhaps, most effective pharmacological intervention for tobacco dependence is varenicline (Cahill et al. 2012, 2013; Rollema et al. 2007b, 2010). This compound is pharmacologically related to a naturally occurring compound, cytisine, that has been used as an aid to smoking cessation for some years, mostly in Central and Eastern Europe (Hajek et al. 2013). Like cytisine, varenicline is a high-affinity partial agonist at the α4β2 subtype of nicotinic receptor but full agonist at the α7 subtype (Mihalak et al. 2006; Rollema et al. 2007b). Experimental animals can be trained to self-administer varenicline and injections of the low doses of the drug enhance brain stimulation reward (Paterson et al. 2010; Rollema et al. 2007a; Spiller et al. 2009). Studies with selective nicotinic receptor antagonists suggest that these effects are mediated by α4β2 receptors. Injections of varenicline also stimulate DA overflow in the accumbal shell although the maximum size of the response is only approximately 50 % of that evoked by nicotine injections (Rollema et al. 2007a). Studies with genetically modified animals suggest that this response depends upon the stimulation of α4β2 receptors in the VTA (Reperant et al. 2010). Thus, reflecting its partial agonist properties at α4β2 receptors, varenicline has many of the principal psychopharmacology properties of nicotine, but with lower efficacy. Pretreating animals with varenicline, however, attenuates the effects of a nicotine injection on DA overflow in the accumbal shell (Rollema et al. 2007a). Furthermore, pretreatment with varenicline also inhibits nicotine self-administration and attenuates the reinstatement of nicotine self-administration evoked by a priming dose of nicotine (O'Connor et al. 2010; Rollema et al. 2007a; Wouda et al. 2011). These responses to varenicline have also been attributed to its partial agonist properties at α4β2 receptors. Varenicline, however, does not inhibit the reinstatement of nicotine-seeking behavior evoked by exposure to a conditioned cue paired with nicotine (O'Connor et al. 2010; Wouda et al. 2011), data that imply that its effects on behavior are related specifically to its interaction with neuronal nicotinic receptors.

8 Conclusions

The data discussed in this chapter have shown that the effects of nicotine on the DA projections to the nucleus accumbens are very similar to those of other drugs that are addictive in human beings, especially the psychostimulants, and that the effects are likely to play an important role in the nicotine dependence. The nucleus accumbens can be seen as the fulcrum of a complex neural system that integrates the neural inputs which influence goal direct behavior and economic decision making (Floresco et al. 2008; Salamone et al. 2012; Schultz 2004; Sesack and Grace 2010). Many of these neural inputs employ glutamatergic and GABAergic neurons to form "hard-wired" connections between the various anatomical loci that are implicated in the regulation of behavior and the nucleus accumbens. The DA projections to the accumbens may best be seen as having a modulatory influence that permits cues associated the availability or presentation of rewards to optimize reward outcomes (Sesack and Grace 2010). Nicotine, like other psychostimulant drugs of dependence, has a direct and marked effect on DA release in the nucleus accumbens. This has significant effects on the ability of habitual users to relate their behavior appropriately to drug rewards and to stimuli associated with the availability of the drug. While it is clear that nicotine acts on the DA projections to the accumbens in a manner similar to other drugs of dependence, it is also clear that nicotine dependence is relatively weak and cannot explain the powerful addiction to tobacco experienced by most habitual smokers.

This chapter has focused on the hypothesis that this conundrum may be explained by the fact that the interactions between nicotine and sensory and spatial contextual stimuli play a pivotal role in the development of nicotine-seeking behavior and the development of the addiction to tobacco. The complementary roles that the shell and the core subdivisions of the accumbens play in mediating the primary reinforcing properties of nicotine, and the secondary reinforcing properties of conditioned stimuli associated with delivery of the drug have been emphasized. The evidence suggests that the reinforcing properties of nicotine are less associated with the "rewarding" properties of the drug itself but reflect its ability to enhance the effects that non-drug rewards exert over behavior. The available evidence suggests that stimuli associated with the availability of nicotine and its administration are important mediators of nicotine-seeking behaviors. It seems reasonable to suggest that existing pharmacotherapies for tobacco addiction are, at best, only moderately successful probably reflect the fact that they do not address this crucial aspect of the dependence. Thus, novel therapies for tobacco dependence and relapse prevention need to focus more on attenuating the effects of these stimuli.

The evidence that mGluR5 receptor antagonists attenuate both the reinforcing properties of nicotine *per se* and the role of conditioned contextual stimuli in eliciting nicotine-seeking behavior points to a potential novel approach to this problem. MPEP has too many off-target effects and too short a half-life to be viable as a therapy for dependence. However, other negative allosteric modulators of the receptor are now being evaluated for the potential safety and efficacy as treatments for psychostimulant dependence in humans (Keck et al. 2013). This type of drug, therefore, may provide a novel approach to the treatment of tobacco dependence. Interestingly, negative allosteric modulators of the mGluR5 receptor are also showing promise in the treatment of psychiatric disorders, such as depression and anxiety, that are significant vulnerability factors for tobacco dependence (Hashimoto et al. 2013). Thus, the potential efficacy of mGluR5 allosteric modulators in the treatment of tobacco dependence may extend further than their effects on contextual conditioning to nicotine.

This chapter has focused particularly on the results of laboratory-based experimental studies that have sought to establish the role of mesolimbic DA in the development of dependence on tobacco smoke. However, the concepts discussed in the chapter may also be relevant to the psychopharmacology of other recreational nicotine delivery systems that many people in the field of public health hope will replace conventional cigarettes. Recent years have seen the introduction of electronic cigarettes (e-cigarettes). These "cigarettes," better described as electronic nicotine delivery systems, deliver nicotine as a heated vapor (Caponnetto et al. 2012). They have the clear advantage that the user does not inhale the toxic pyrolysis products produced by burning tobacco and substantially reduces any potential harm to others caused by exposure to second-hand smoke. While, at present, e-cigarettes are used primarily by people who are current or previous smokers of conventional cigarettes, they are marketed as an alternative and safer means of using nicotine recreationally which may also be adopted by people who have not previously smoked (Dockrell et al. 2013). Little is currently known of the potential abuse liability of e-cigarettes (Fagerstrom and Eissenberg 2012). Recent studies, however, suggest that, when experienced users inhale the vapor from e-cigarettes, they can achieve elevations of plasma nicotine that are rapid and comparable to those achieved by smoking a conventional cigarette (Etter and Bullen 2011; Vansickel and Eissenberg 2013). It is reasonable, therefore, to speculate that these nicotine delivery systems may have similar effects on mesoaccumbens DA release as those evoked by smoking a conventional cigarette. Moreover, while e-cigarette users may not be exposed to the complex sensory cues associated with inhaling tobacco smoke, habitual use of the devices will potentially expose the user to sensory cues associate with delivery of the vapor and to distal contextual cues associated with environments in which the device is used (Fagerstrom and Eissenberg 2012). Thus, it seems reasonable to suggest that although e-cigarette use is significantly less harmful than smoking regular cigarettes, their vapor is likely to retain some of the addictive potential of tobacco smoke. This is an issue that may need to be considered by regulatory authorities.

Acknowledgments The studies reported in this chapter from the author's laboratory were performed with the aid of grants from the Wellcome Trust and Cancer Research UK.

References

Alderson HL, Latimer MP, Winn P (2006) Intravenous self-administration of nicotine is altered by lesions of the posterior, but not anterior, pedunculopontine tegmental nucleus. Eur J Neurosci 23:2169–2175

Aragona BJ, Cleaveland NA, Stuber GD, Day JJ, Carelli RM, Wightman RM (2008) Preferential enhancement of dopamine transmission within the nucleus accumbens shell by cocaine is attributable to a direct increase in phasic dopamine release events. J Neurosci 28:8821–8831

Balfour DJ (2004) The neurobiology of tobacco dependence: a preclinical perspective on the role of the dopamine projections to the nucleus accumbens. Nicotine Tob Res 6:899–912

Balfour DJ (2009) The neuronal pathways mediating the behavioral and addictive properties of nicotine. Hand Exp Pharmacol 192:209–233

Balfour DJ, Birrell CE, Moran RJ, Benwell ME (1996) Effects of acute D-CPPene on mesoaccumbens dopamine responses to nicotine in the rat. Eur J Pharmacol 316:153–156

Balfour DJ, Benwell ME, Birrell CE, Kelly RJ, Al-Aloul M (1998) Sensitization of the mesoaccumbens dopamine response to nicotine. Pharmacol Biochem Behav 59:1021–1030

Balfour DJ, Wright AE, Benwell ME, Birrell CE (2000) The putative role of extra-synaptic mesolimbic dopamine in the neurobiology of nicotine dependence. Behav Brain Res 113:73–83

Bassareo V, De Luca MA, Di Chiara G (2007) Differential impact of pavlovian drug conditioned stimuli on in vivo dopamine transmission in the rat accumbens shell and core and in the prefrontal cortex. Psychopharmacology 191:689–703

Beard E, McNeill A, Aveyard P, Fidler J, Michie S, West R (2013) Association between use of nicotine replacement therapy for harm reduction and smoking cessation: a prospective study of English smokers. Tob Control 22:118–122

Benwell ME, Balfour DJ (1992) The effects of acute and repeated nicotine treatment on nucleus accumbens dopamine and locomotor activity. Br J Pharmacol 105:849–856

Benwell ME, Balfour DJ (1997) Regional variation in the effects of nicotine on catecholamine overflow in rat brain. Eur J Pharmacol 325:13–20

Benwell ME, Balfour DJ, Birrell CE (1995) Desensitization of the nicotine-induced mesolimbic dopamine responses during constant infusion with nicotine. Br J Pharmacol 114:454–460

Benwell ME, Holtom PE, Moran RJ, Balfour DJ (1996) Neurochemical and behavioural interactions between ibogaine and nicotine in the rat. Br J Pharmacol 117:743–749

Benowitz NC, Porcheth, Jacob P (1990) Pharmacokinetics metabolism and pharmacodynamics of nicotine. In: Wonnacott S, Russell MAH, Stolerman IP (eds). Niconne psychopharmacology: molecular cellular and behavioural aspects. Oxford University Press, Oxford, pp 112–157

Bespalov AY, Dravolina OA, Sukhanov I, Zakharova E, Blokhina E, Zvartau E, Danysz W, van Heeke G, Markou A (2005) Metabotropic glutamate receptor (mGluR5) antagonist MPEP attenuated cue- and schedule-induced reinstatement of nicotine self-administration behavior in rats. Neuropharmacology 491:167–178

Bevins RA, Palmatier MI (2003) Nicotine-conditioned locomotor sensitization in rats: assessment of the US-preexposure effect. Behav Brain Res 143:65–74

Birrell CE, Balfour DJ (1998) The influence of nicotine pretreatment on mesoaccumbens dopamine overflow and locomotor responses to D-amphetamine. Psychopharmacology 140:142–149

Bossert JM, Poles GC, Wihbey KA, Koya E, Shaham Y (2007) Differential effects of blockade of dopamine D1-family receptors in nucleus accumbens core or shell on reinstatement of heroin seeking induced by contextual and discrete cues. J Neurosci 27:12655–12663

Boye SM, Grant RJ, Clarke PB (2001) Disruption of dopaminergic neurotransmission in nucleus accumbens core inhibits the locomotor stimulant effects of nicotine and D-amphetamine in rats. Neuropharmacology 40:792–805

Bozarth MA, Wise RA (1984) Anatomically distinct opiate receptor fields mediate reward and physical dependence. Science 224:516–517

Brauer LH, Behm FM, Lane JD, Westman EC, Perkins C, Rose JE (2001) Individual differences in smoking reward from de-nicotinized cigarettes. Nicotine Tob Res 3:101–109

Brazell MP, Mitchell SN, Joseph MH, Gray JA (1990) Acute administration of nicotine increases the in vivo extracellular levels of dopamine, 3,4-dihydroxyphenylacetic acid and ascorbic acid preferentially in the nucleus accumbens of the rat: comparison with caudate-putamen. Neuropharmacology 29:1177–1185

Brody AL (2006) Functional brain imaging of tobacco use and dependence. J Psychiatr Res 40:404–418

Brody AL, Olmstead RE, London ED, Farahi J, Meyer JH, Grossman P, Lee GS, Huang J, Hahn EL, Mandelkern MA (2004) Smoking-induced ventral striatum dopamine release. Am J Psychiatry 161:1211–1218

Brody AL, London ED, Olmstead RE, Allen-Martinez Z, Shulenberger S, Costello MR, Abrams AL, Scheibal D, Farahi J, Shoptaw S, Mandelkern MA (2010) Smoking-induced change in intrasynaptic dopamine concentration: effect of treatment for tobacco dependence. Psychiatry Res 183:218–224

Brower VG, Fu Y, Matta SG, Sharp BM (2002) Rat strain differences in nicotine self-administration using an unlimited access paradigm. Brain Res 930:12–20

Bruijnzeel AW, Markou A (2004) Adaptations in cholinergic transmission in the ventral tegmental area associated with the affective signs of nicotine withdrawal in rats. Neuropharmacology 47:572–579

Cadoni C, Di Chiara G (1999) Reciprocal changes in dopamine responsiveness in the nucleus accumbens shell and core and in the dorsal caudate-putamen in rats sensitized to morphine. Neuroscience 90:447–455

Cadoni C, Di Chiara G (2000) Differential changes in accumbens shell and core dopamine in behavioral sensitization to nicotine. Eur J Pharmacol 387:R23–R25

Cadoni C, Solinas M, Di Chiara G (2000) Psychostimulant sensitization: differential changes in accumbal shell and core dopamine. Eur J Pharmacol 388:69–76

Cadoni C, Muto T, Di Chiara G (2009) Nicotine differentially affects dopamine transmission in the nucleus accumbens shell and core of Lewis and Fischer 344 rats. Neuropharmacology 57:496–501

Caggiula AR, Donny EC, White AR, Chaudhri N, Booth S, Gharib MA, Hoffman A, Perkins KA, Sved AF (2001) Cue dependency of nicotine self-administration and smoking. Pharmacol Biochem Behav 70:515–530

Caggiula AR, Donny EC, Chaudhri N, Perkins KA, Evans-Martin FF, Sved AF (2002) Importance of nonpharmacological factors in nicotine self-administration. Physiol Behav 77:683–687

Cahill K, Stead LF, Lancaster T (2012) Nicotine receptor partial agonists for smoking cessation. The Cochrane database of systematic reviews 4: CD006103

Cahill K, Stevens S, Perera R, Lancaster T (2013) Pharmacological interventions for smoking cessation: an overview and network meta-analysis. Cochrane Database Syst Rev 5:CD009329

Cannon CM, Palmiter RD (2003) Reward without dopamine. J Neurosci 23:10827–10831

Caponnetto P, Campagna D, Papale G, Russo C, Polosa R (2012) The emerging phenomenon of electronic cigarettes. Exp Rev Resp Med 6:63–74

Carboni E, Bortone L, Giua C, Di Chiara G (2000) Dissociation of physical abstinence signs from changes in extracellular dopamine in the nucleus accumbens and in the prefrontal cortex of nicotine dependent rats. Drug Alcohol Depend 58:93–102

Chaudhri N, Caggiula AR, Donny EC, Palmatier MI, Liu X, Sved AF (2006) Complex interactions between nicotine and nonpharmacological stimuli reveal multiple roles for nicotine in reinforcement. Psychopharmacology 184:353–366

Chaudhri N, Sahuque LL, Schairer WW, Janak PH (2010) Separable roles of the nucleus accumbens core and shell in context—and cue-induced alcohol-seeking. Neuropsychopharmacology 35:783–791

Chergui K, Charlety PJ, Akaoka H, Saunier CF, Brunet JL, Buda M, Svensson TH, Chouvet G (1993) Tonic activation of NMDA receptors causes spontaneous burst discharge of rat midbrain dopamine neurons in vivo. Eur J Neurosci 5:137–144

Clarke PB, Kumar R (1983) The effects of nicotine on locomotor activity in non-tolerant and tolerant rats. Br J Pharmacol 78:329–337

Clarke PB, Fu DS, Jakubovic A, Fibiger HC (1988) Evidence that mesolimbic dopaminergic activation underlies the locomotor stimulant action of nicotine in rats. J Pharmacol Exp Ther 246:701–708

Coen KM, Adamson KL, Corrigall WA (2009) Medication-related pharmacological manipulations of nicotine self-administration in the rat maintained on fixed- and progressive-ratio schedules of reinforcement. Psychopharmacology 201:557–568

Cohen C, Perrault G, Voltz C, Steinberg R, Soubrie P (2002) SR141716, a central cannabinoid (CB(1)) receptor antagonist, blocks the motivational and dopamine-releasing effects of nicotine in rats. Behav Pharmacol 13:451–463

Cohen C, Kodas E, Griebel G (2005) CB1 receptor antagonists for the treatment of nicotine addiction. Pharmacol Biochem Behav 81:387–395

Corrigall WA, Coen KM (1989) Nicotine maintains robust self-administration in rats on a limited-access schedule. Psychopharmacology 99:473–478

Corrigall WA, Franklin KB, Coen KM, Clarke PB (1992) The mesolimbic dopaminergic system is implicated in the reinforcing effects of nicotine. Psychopharmacology 107:285–289

Corrigall WA, Coen KM, Adamson KL (1994) Self-administered nicotine activates the mesolimbic dopamine system through the ventral tegmental area. Brain Res 653:278–284

Corrigall WA, Coen KM, Zhang J, Adamson KL (2001) GABA mechanisms in the pedunculopontine tegmental nucleus influence particular aspects of nicotine self-administration selectively in the rat. Psychopharmacology 158:190–197

Corrigall WA, Coen KM, Zhang J, Adamson L (2002) Pharmacological manipulations of the pedunculopontine tegmental nucleus in the rat reduce self-administration of both nicotine and cocaine. Psychopharmacology 160:198–205

Crippens D, Robinson TE (1994) Withdrawal from morphine or amphetamine: different effects on dopamine in the ventral-medial striatum studied with microdialysis. Brain Res 650:56–62

Crippens D, Camp DM, Robinson TE (1993) Basal extracellular dopamine in the nucleus accumbens during amphetamine withdrawal: a 'no net flux' microdialysis study. Neurosci Lett 164:145–148

Crombag HS, Bossert JM, Koya E, Shaham Y (2008) Context-induced relapse to drug seeking: a review. Philos Trans R Soc Biol Sci 363:3233–3243

Cryan JF, Hoyer D, Markou A (2003) Withdrawal from chronic amphetamine induces depressive-like behavioral effects in rodents. Biol Psychiatry 54:49–58

Dackis CA, Gold MS (1985) New concepts in cocaine addiction: the dopamine depletion hypothesis. Neurosci Biobehav Rev 9:469–477

Dagher A, Bleicher C, Aston JA, Gunn RN, Clarke PB, Cumming P (2001) Reduced dopamine D1 receptor binding in the ventral striatum of cigarette smokers. Synapse 42:48–53

Damsma G, Day J, Fibiger HC (1989) Lack of tolerance to nicotine-induced dopamine release in the nucleus accumbens. Eur J Pharmacol 168:363–368

Dawe S, Gerada C, Russell MA, Gray JA (1995) Nicotine intake in smokers increases following a single dose of haloperidol. Psychopharmacology 117:110–115

Di Chiara G (1995) The role of dopamine in drug abuse viewed from the perspective of its role in motivation. Drug Alcohol Depend 38:95–137

Di Chiara G (1999) Drug addiction as dopamine-dependent associative learning disorder. Eur J Pharmacol 375:13–30

Di Chiara G (2000) Role of dopamine in the behavioural actions of nicotine related to addiction. Eur J Pharmacol 393:295–314

Di Chiara G (2002) Nucleus accumbens shell and core dopamine: differential role in behavior and addiction. Behav Brain Res 137:75–114

Di Chiara G, Bassareo V (2007) Reward system and addiction: what dopamine does and doesn't do. Curr Opin Pharmacol 7:69–76

Di Chiara G, Imperato A (1988) Drugs abused by humans preferentially increase synaptic dopamine concentrations in the mesolimbic system of freely moving rats. Proc Natl Acad Sci USA 85:5274–5278

Di Chiara G, Bassareo V, Fenu S, De Luca MA, Spina L, Cadoni C, Acquas E, Carboni E, Valentini V, Lecca D (2004) Dopamine and drug addiction: the nucleus accumbens shell connection. Neuropharmacology 47:227–241

Diergaarde L, de Vries W, Raaso H, Schoffelmeer AN, De Vries TJ (2008) Contextual renewal of nicotine seeking in rats and its suppression by the cannabinoid-1 receptor antagonist Rimonabant (SR141716A). Neuropharmacology 55:712–716

Dockrell M, Morrison R, Bauld L, McNeill A (2013) E-cigarettes: prevalence and attitudes in Great Britain. Nicotine Tob Res 15:1737–1744

Donny EC, Chaudhri N, Caggiula AR, Evans-Martin FF, Booth S, Gharib MA, Clements LA, Sved AF (2003) Operant responding for a visual reinforcer in rats is enhanced by noncontingent nicotine: implications for nicotine self-administration and reinforcement. Psychopharmacology 169:68–76

D'Souza MS, Markou A (2010) Neural substrates of psychostimulant withdrawal-induced anhedonia. Curr Top Behav Neurosci 3:119–178

D'Souza MS, Markou A (2011) Metabotropic glutamate receptor 5 antagonist 2-methyl-6-(phenylethynyl)pyridine (MPEP) microinfusions into the nucleus accumbens shell or ventral tegmental area attenuate the reinforcing effects of nicotine in rats. Neuropharmacology 61:1399–1405

D'Souza MS, Markou A (2014) Differential role of N-methyl-D-aspartate receptor-mediated glutamate transmission in the nucleus accumbens shell and core in nicotine seeking in rats. Eur J Neurosci

Epping-Jordan MP, Watkins SS, Koob GF, Markou A (1998) Dramatic decreases in brain reward function during nicotine withdrawal. Nature 393:76–79

Etter JF, Bullen C (2011) Electronic cigarette: users profile, utilization, satisfaction and perceived efficacy. Addiction 106:2017–2028

Fagen ZM, Mansvelder HD, Keath JR, McGehee DS (2003) Short- and long-term modulation of synaptic inputs to brain reward areas by nicotine. Ann NY Acad Sci 1003:185–195

Fagerstrom K, Eissenberg T (2012) Dependence on tobacco and nicotine products: a case for product-specific assessment. Nicotine Tob Res 14:1382–1390

Farquhar MJ, Latimer MP, Winn P (2012) Nicotine self-administered directly into the VTA by rats is weakly reinforcing but has strong reinforcement enhancing properties. Psychopharmacology 220:43–54

Floresco SB (2007) Dopaminergic regulation of limbic-striatal interplay. Journal Psychiatry Neurosci 32:400–411

Floresco SB, Todd CL, Grace AA (2001) Glutamatergic afferents from the hippocampus to the nucleus accumbens regulate activity of ventral tegmental area dopamine neurons. J Neurosci 21:4915–4922

Floresco SB, West AR, Ash B, Moore H, Grace AA (2003) Afferent modulation of dopamine neuron firing differentially regulates tonic and phasic dopamine transmission. Nat Neurosci 6:968–973

Floresco SB, St Onge JR, Ghods-Sharifi S, Winstanley CA (2008) Cortico-limbic-striatal circuits subserving different forms of cost-benefit decision making. Cogn Affect Behav Neurosci 8:375–389

Fowler JS, Logan J, Wang GJ, Volkow ND (2003) Monoamine oxidase and cigarette smoking. Neurotoxicology 24:75–82

Fu Y, Matta SG, Gao W, Brower VG, Sharp BM (2000) Systemic nicotine stimulates dopamine release in nucleus accumbens: re-evaluation of the role of N-methyl-D-aspartate receptors in the ventral tegmental area. J Pharmacol Exp Ther 294:458–465

Fuxe K, Dahlstrom AB, Jonsson G, Marcellino D, Guescini M, Dam M, Manger P, Agnati L (2010) The discovery of central monoamine neurons gave volume transmission to the wired brain. Prog Neurobiol 90:82–100

Garris PA, Wightman RM (1994) Different kinetics govern dopaminergic transmission in the amygdala, prefrontal cortex, and striatum: an in vivo voltammetric study. J Neurosci 14:442–450

Gerrits MA, Van Ree JM (1996) Effect of nucleus accumbens dopamine depletion on motivational aspects involved in initiation of cocaine and heroin self-administration in rats. Brain Res 713:114–124

Glick SD, Maisonneuve IM (2000) Development of novel medications for drug addiction. The legacy of an African shrub. Ann NY Acad Sci 909:88–103

Gold MS, Dackis CA (1984) New insights and treatments: opiate withdrawal and cocaine addiction. Clin Ther 7:6–21

Grace AA (1991) Phasic versus tonic dopamine release and the modulation of dopamine system responsivity: a hypothesis for the etiology of schizophrenia. Neuroscience 41:1–24

Grace AA (2000) The tonic/phasic model of dopamine system regulation and its implications for understanding alcohol and psychostimulant craving. Addiction 95:S119–S128

Grace AA, Bunney BS (1984a) The control of firing pattern in nigral dopamine neurons: burst firing. J Neurosci 4:2877–2890

Grace AA, Bunney BS (1984b) The control of firing pattern in nigral dopamine neurons: single spike firing. J Neurosci 4:2866–2876

Grace AA, Floresco SB, Goto Y, Lodge DJ (2007) Regulation of firing of dopaminergic neurons and control of goal-directed behaviors. Trends Neurosci 30:220–227

Grenhoff J, Aston-Jones G, Svensson TH (1986) Nicotinic effects on the firing pattern of midbrain dopamine neurons. Acta Physiol Scand 128:351–358

Grieder TE, George O, Tan H, George SR, Le Foll B, Laviolette SR, van der Kooy D (2012) Phasic D1 and tonic D2 dopamine receptor signaling double dissociate the motivational effects of acute nicotine and chronic nicotine withdrawal. Proc Natl Acad Sci USA 109:3101–3106

Guillem K, Vouillac C, Azar MR, Parsons LH, Koob GF, Cador M, Stinus L (2005) Monoamine oxidase inhibition dramatically increases the motivation to self-administer nicotine in rats. J Neurosci 25:8593–8600

Guillem K, Vouillac C, Azar MR, Parsons LH, Koob GF, Cador M, Stinus L (2006) Monoamine oxidase A rather than monoamine oxidase B inhibition increases nicotine reinforcement in rats. Eur J Neurosci 24:3532–3540

Hajek P, McRobbie H, Myers K (2013) Efficacy of cytisine in helping smokers quit: systematic review and meta-analysis. Thorax 68:1037–1042

Hashimoto K, Malchow B, Falkai P, Schmitt A (2013) Glutamate modulators as potential therapeutic drugs in schizophrenia and affective disorders. Eur Arch Psychiatry Clin Neurosci 263:367–377

Heien ML, Wightman RM (2006) Phasic dopamine signaling during behavior, reward, and disease states. CNS Neurol Disord: Drug Targets 5:99–108

Heimer L, Zahm DS, Churchill L, Kalivas PW, Wohltmann C (1991) Specificity in the projection patterns of accumbal core and shell in the rat. Neuroscience 41:89–125

Hersch SM, Yi H, Heilman CJ, Edwards RH, Levey AI (1997) Subcellular localization and molecular topology of the dopamine transporter in the striatum and substantia nigra. J Comp Neurol 388:211–227

Hildebrand BE, Svensson TH (2000) Intraaccumbal mecamylamine infusion does not affect dopamine output in the nucleus accumbens of chronically nicotine-treated rats. J Neural Transm 107:861–872

Hildebrand BE, Nomikos GG, Bondjers C, Nisell M, Svensson TH (1997) Behavioral manifestations of the nicotine abstinence syndrome in the rat: peripheral versus central mechanisms. Psychopharmacology 129:348–356

Hildebrand BE, Nomikos GG, Hertel P, Schilstrom B, Svensson TH (1998) Reduced dopamine output in the nucleus accumbens but not in the medial prefrontal cortex in rats displaying a mecamylamine-precipitated nicotine withdrawal syndrome. Brain Res 779:214–225

Hollander JA, Carelli RM (2007) Cocaine-associated stimuli increase cocaine seeking and activate accumbens core neurons after abstinence. J Neurosci 27:3535–3539

Ikemoto S (2003) Involvement of the olfactory tubercle in cocaine reward: intracranial self-administration studies. J Neurosci 23:9305–9311

Imperato A, Mulas A, Di Chiara G (1986) Nicotine preferentially stimulates dopamine release in the limbic system of freely moving rats. Eur J Pharmacol 132:337–338

Ito R, Dalley JW, Howes SR, Robbins TW, Everitt BJ (2000) Dissociation in conditioned dopamine release in the nucleus accumbens core and shell in response to cocaine cues and during cocaine-seeking behavior in rats. J Neurosci 20:7489–7495

Ito R, Dalley JW, Robbins TW, Everitt BJ (2002) Dopamine release in the dorsal striatum during cocaine-seeking behavior under the control of a drug-associated cue. J Neurosci 22:6247–6253

Ito R, Robbins TW, Everitt BJ (2004) Differential control over cocaine-seeking behavior by nucleus accumbens core and shell. Nat Neurosci 7:389–397

Ito R, Robbins TW, Pennartz CM, Everitt BJ (2008) Functional interaction between the hippocampus and nucleus accumbens shell is necessary for the acquisition of appetitive spatial context conditioning. J Neurosci 28:6950–6959

Iyaniwura TT, Wright AE, Balfour DJ (2001) Evidence that mesoaccumbens dopamine and locomotor responses to nicotine in the rat are influenced by pretreatment dose and strain. Psychopharmacology 158:73–79

Johnston AJ, Ascher J, Leadbetter R, Schmith VD, Patel DK, Durcan M, Bentley B (2002) Pharmacokinetic optimisation of sustained-release bupropion for smoking cessation. Drugs 62:11–24

Jones SR, O'Dell SJ, Marshall JF, Wightman RM (1996) Functional and anatomical evidence for different dopamine dynamics in the core and shell of the nucleus accumbens in slices of rat brain. Synapse 23:224–231

Keck TM, Yang HJ, Bi GH, Huang Y, Zhang HY, Srivastava R, Gardner EL, Newman AH, Xi ZX (2013) Fenobam sulfate inhibits cocaine-taking and cocaine-seeking behavior in rats: implications for addiction treatment in humans. Psychopharmacology 229:253–265

Kenny PJ, Markou A (2001) Neurobiology of the nicotine withdrawal syndrome. Pharmacol Biochem Behav 70:531–549

Kenny PJ, Paterson NE, Boutrel B, Semenova S, Harrison AA, Gasparini F, Koob GF, Skoubis PD, Markou A (2003) Metabotropic glutamate 5 receptor antagonist MPEP decreased nicotine and cocaine self-administration but not nicotine and cocaine-induced facilitation of brain reward function in rats. Ann NY Acad Sci 1003:415–418

Kenny PJ, Chartoff E, Roberto M, Carlezon WA Jr, Markou A (2009) NMDA receptors regulate nicotine-enhanced brain reward function and intravenous nicotine self-administration: role of the ventral tegmental area and central nucleus of the amygdala. Neuropsychopharmacology 34:266–281

Kraiczi H, Hansson A, Perfekt R (2011) Single-dose pharmacokinetics of nicotine when given with a novel mouth spray for nicotine replacement therapy. Nicotine Tob Res 13:1176–1182

Kyerematen GA, Taylor LH, deBethizy JD, Vesell ES (1988) Pharmacokinetics of nicotine and 12 metabolites in the rat. Application of a new radiometric high performance liquid chromatography assay. Drug Metab Dispos 16:125–129

Lanca AJ, Adamson KL, Coen KM, Chow BL, Corrigall WA (2000) The pedunculopontine tegmental nucleus and the role of cholinergic neurons in nicotine self-administration in the rat: a correlative neuroanatomical and behavioral study. Neuroscience 96:735–742

Le Foll B, Goldberg SR (2005) Control of the reinforcing effects of nicotine by associated environmental stimuli in animals and humans. Trends Pharmacol Sci 26:287–293

Lecca D, Cacciapaglia F, Valentini V, Gronli J, Spiga S, Di Chiara G (2006) Preferential increase of extracellular dopamine in the rat nucleus accumbens shell as compared to that in the core during acquisition and maintenance of intravenous nicotine self-administration. Psychopharmacology 184:435–446

Lecca D, Cacciapaglia F, Valentini V, Acquas E, Di Chiara G (2007a) Differential neurochemical and behavioral adaptation to cocaine after response contingent and noncontingent exposure in the rat. Psychopharmacology 191:653–667

Lecca D, Valentini V, Cacciapaglia F, Acquas E, Di Chiara G (2007b) Reciprocal effects of response contingent and noncontingent intravenous heroin on in vivo nucleus accumbens shell versus core dopamine in the rat: a repeated sampling microdialysis study. Psychopharmacology 194:103–116

Legault M, Wise RA (1999) Injections of N-methyl-D-aspartate into the ventral hippocampus increase extracellular dopamine in the ventral tegmental area and nucleus accumbens. Synapse 31:241–249

LeSage MG, Keyler DE, Collins G, Pentel PR (2003) Effects of continuous nicotine infusion on nicotine self-administration in rats: relationship between continuously infused and self-administered nicotine doses and serum concentrations. Psychopharmacology 170:278–286

LeSage MG, Burroughs D, Dufek M, Keyler DE, Pentel PR (2004) Reinstatement of nicotine self-administration in rats by presentation of nicotine-paired stimuli, but not nicotine priming. Pharmacol Biochem Behav 79:507–513

Li W, Doyon WM, Dani JA (2011) Acute in vivo nicotine administration enhances synchrony among dopamine neurons. Biochem Pharmacol 82:977–983

Livingstone PD, Wonnacott S (2009) Nicotinic acetylcholine receptors and the ascending dopamine pathways. Biochem Pharmacol 78:744–755

Lodge DJ, Grace AA (2006a) The hippocampus modulates dopamine neuron responsivity by regulating the intensity of phasic neuron activation. Neuropsychopharmacology 31:1356–1361

Lodge DJ, Grace AA (2006b) The laterodorsal tegmentum is essential for burst firing of ventral tegmental area dopamine neurons. Proc Natl Acad Sci USA 103:5167–5172

Louis M, Clarke PB (1998) Effect of ventral tegmental 6-hydroxydopamine lesions on the locomotor stimulant action of nicotine in rats. Neuropharmacology 37:1503–1513

Lu XY, Ghasemzadeh MB, Kalivas PW (1999) Expression of glutamate receptor subunit/subtype messenger RNAS for NMDAR1, GLuR1, GLuR2 and mGLuR5 by accumbal projection neurons. Brain Res Mol Brain Res 63:287–296

Lyness WH, Friedle NM, Moore KE (1979) Destruction of dopaminergic nerve terminals in nucleus accumbens: effect on d-amphetamine self-administration. Pharmacol Biochem Behav 11:553–556

Malin DH (2001) Nicotine dependence: studies with a laboratory model. Pharmacol Biochem Behav 70:551–559

Malin DH, Goyarzu P (2009) Rodent models of nicotine withdrawal syndrome. Handb Exp Pharmacol 192:401–434

Malin DH, Lake JR, Newlin-Maultsby P, Roberts LK, Lanier JG, Carter VA, Cunningham JS, Wilson OB (1992) Rodent model of nicotine abstinence syndrome. Pharmacol Biochem Behav 43:779–784

Malin DH, Lake JR, Carter VA, Cunningham JS, Hebert KM, Conrad DL, Wilson OB (1994) The nicotinic antagonist mecamylamine precipitates nicotine abstinence syndrome in the rat. Psychopharmacology 115:180–184

Malin DH, Lake JR, Smith TD, Khambati HN, Meyers-Paal RL, Montellano AL, Jennings RE, Erwin DS, Presley SE, Perales BA (2006) Bupropion attenuates nicotine abstinence syndrome in the rat. Psychopharmacology 184:494–503

Mameli-Engvall M, Evrard A, Pons S, Maskos U, Svensson TH, Changeux JP, Faure P (2006) Hierarchical control of dopamine neuron-firing patterns by nicotinic receptors. Neuron 50:911–921

Mansvelder HD, McGehee DS (2000) Long-term potentiation of excitatory inputs to brain reward areas by nicotine. Neuron 27:349–357

Mansvelder HD, McGehee DS (2002) Cellular and synaptic mechanisms of nicotine addiction. J Neurobiol 53:606–617

Mansvelder HD, Keath JR, McGehee DS (2002) Synaptic mechanisms underlie nicotine-induced excitability of brain reward areas. Neuron 33:905–919

Markou A (2008) Review. Neurobiology of nicotine dependence. Philos Trans R Soc Lond B Biol Sci 363:3159–3168

Markou A, Koob GF (1991) Postcocaine anhedonia. An animal model of cocaine withdrawal. Neuropsychopharmacology 4:17–26

Mascia P, Pistis M, Justinova Z, Panlilio LV, Luchicchi A, Lecca S, Scherma M, Fratta W, Fadda P, Barnes C, Redhi GH, Yasar S, Le Foll B, Tanda G, Piomelli D, Goldberg SR (2011) Blockade of nicotine reward and reinstatement by activation of alpha-type peroxisome proliferator-activated receptors. Biol Psychiatry 69:633–641

Mereu G, Yoon KW, Boi V, Gessa GL, Naes L, Westfall TC (1987) Preferential stimulation of ventral tegmental area dopaminergic neurons by nicotine. Eur J Pharmacol 141:395–399

Mihalak KB, Carroll FI, Luetje CW (2006) Varenicline is a partial agonist at alpha4beta2 and a full agonist at alpha7 neuronal nicotinic receptors. Mol Pharmacol 70:801–805

Nirenberg MJ, Chan J, Pohorille A, Vaughan RA, Uhl GR, Kuhar MJ, Pickel VM (1997) The dopamine transporter: comparative ultrastructure of dopaminergic axons in limbic and motor compartments of the nucleus accumbens. J Neurosci 17:6899–6907

Nisell M, Nomikos GG, Svensson TH (1994a) Infusion of nicotine in the ventral tegmental area or the nucleus accumbens of the rat differentially affects accumbal dopamine release. Pharmacol Toxicol 75:348–352

Nisell M, Nomikos GG, Svensson TH (1994b) Systemic nicotine-induced dopamine release in the rat nucleus accumbens is regulated by nicotinic receptors in the ventral tegmental area. Synapse 16:36–44

Nunes EJ, Randall PA, Podurgiel S, Correa M, Salamone JD (2013) Nucleus accumbens neurotransmission and effort-related choice behavior in food motivation: effects of drugs acting on dopamine, adenosine, and muscarinic acetylcholine receptors. Neurosci Biobehav Rev 37:2015–2025

O'Connor EC, Parker D, Rollema H, Mead AN (2010) The alpha4beta2 nicotinic acetylcholine-receptor partial agonist varenicline inhibits both nicotine self-administration following repeated dosing and reinstatement of nicotine seeking in rats. Psychopharmacology 208:365–376

O'Dell LE, Khroyan TV (2009) Rodent models of nicotine reward: what do they tell us about tobacco abuse in humans? Pharmacol Biochem Behav 91:481–488

Owesson-White CA, Roitman MF, Sombers LA, Belle AM, Keithley RB, Peele JL, Carelli RM, Wightman RM (2012) Sources contributing to the average extracellular concentration of dopamine in the nucleus accumbens. J Neurochem 121:252–262

Palmatier MI, Evans-Martin FF, Hoffman A, Caggiula AR, Chaudhri N, Donny EC, Liu X, Booth S, Gharib M, Craven L, Sved AF (2006) Dissociating the primary reinforcing and reinforcement-enhancing effects of nicotine using a rat self-administration paradigm with concurrently available drug and environmental reinforcers. Psychopharmacology 184:391–400

Palmatier MI, Matteson GL, Black JJ, Liu X, Caggiula AR, Craven L, Donny EC, Sved AF (2007) The reinforcement enhancing effects of nicotine depend on the incentive value of non-drug reinforcers and increase with repeated drug injections. Drug Alcohol Depend 89:52–59

Palmatier MI, Liu X, Donny EC, Caggiula AR, Sved AF (2008) Metabotropic glutamate 5 receptor (mGluR5) antagonists decrease nicotine seeking, but do not affect the reinforcement enhancing effects of nicotine. Neuropsychopharmacology 33:2139–2147

Panagis G, Nisell M, Nomikos GG, Chergui K, Svensson TH (1996) Nicotine injections into the ventral tegmental area increase locomotion and Fos-like immunoreactivity in the nucleus accumbens of the rat. Brain Res 730:133–142

Paterson NE, Markou A (2004) Prolonged nicotine dependence associated with extended access to nicotine self-administration in rats. Psychopharmacology 173:64–72

Paterson NE, Markou A (2005) The metabotropic glutamate receptor 5 antagonist MPEP decreased break points for nicotine, cocaine and food in rats. Psychopharmacology 179:255–261

Paterson NE, Markou A (2007) Animal models and treatments for addiction and depression co-morbidity. Neurotox Res 11:1–32

Paterson NE, Semenova S, Gasparini F, Markou A (2003) The mGluR5 antagonist MPEP decreased nicotine self-administration in rats and mice. Psychopharmacology 167:257–264

Paterson NE, Balfour DJ, Markou A (2007) Chronic bupropion attenuated the anhedonic component of nicotine withdrawal in rats via inhibition of dopamine reuptake in the nucleus accumbens shell. Eur J Neurosci 25:3099–3108

Paterson NE, Min W, Hackett A, Lowe D, Hanania T, Caldarone B, Ghavami A (2010) The high-affinity nAChR partial agonists varenicline and sazetidine-A exhibit reinforcing properties in rats. Prog Neuropsychopharmacol Biol Psychiatry 34:1455–1464

Pettit HO, Ettenberg A, Bloom FE, Koob GF (1984) Destruction of dopamine in the nucleus accumbens selectively attenuates cocaine but not heroin self-administration in rats. Psychopharmacology 84:167–173

Phillips PE, Stuber GD, Heien ML, Wightman RM, Carelli RM (2003) Subsecond dopamine release promotes cocaine seeking. Nature 422:614–618

Pidoplichko VI, DeBiasi M, Williams JT, Dani JA (1997) Nicotine activates and desensitizes midbrain dopamine neurons. Nature 390:401–404

Pontieri FE, Tanda G, Orzi F, Di Chiara G (1996) Effects of nicotine on the nucleus accumbens and similarity to those of addictive drugs. Nature 382:255–257

Rada P, Jensen K, Hoebel BG (2001) Effects of nicotine and mecamylamine-induced withdrawal on extracellular dopamine and acetylcholine in the rat nucleus accumbens. Psychopharmacology 157:105–110

Ranaldi R, Roberts DC (1996) Initiation, maintenance and extinction of cocaine self-administration with and without conditioned reward. Psychopharmacology 128:89–96

Rauhut AS, Neugebauer N, Dwoskin LP, Bardo MT (2003) Effect of bupropion on nicotine self-administration in rats. Psychopharmacology 169:1–9

Reavill C, Walther B, Stolerman IP, Testa B (1990) Behavioural and pharmacokinetic studies on nicotine, cytisine and lobeline. Neuropharmacology 29:619–624

Reid MS, Ho LB, Berger SP (1998) Behavioral and neurochemical components of nicotine sensitization following 15-day pretreatment: studies on contextual conditioning. Behav Pharmacol 9:137–148

Reperant C, Pons S, Dufour E, Rollema H, Gardier AM, Maskos U (2010) Effect of the alpha4beta2* nicotinic acetylcholine receptor partial agonist varenicline on dopamine release in beta2 knock-out mice with selective re-expression of the beta2 subunit in the ventral tegmental area. Neuropharmacology 58:346–350

Roberts DC, Koob GF (1982) Disruption of cocaine self-administration following 6-hydroxydopamine lesions of the ventral tegmental area in rats. Pharmacol Biochem Behav 17:901–904

Roberts DC, Corcoran ME, Fibiger HC (1977) On the role of ascending catecholaminergic systems in intravenous self-administration of cocaine. Pharmacol Biochem Behav 6:615–620

Rocha BA, Fumagalli F, Gainetdinov RR, Jones SR, Ator R, Giros B, Miller GW, Caron MG (1998) Cocaine self-administration in dopamine-transporter knockout mice. Nat Neurosci 1:132–137

Rodd-Henricks ZA, McKinzie DL, Li TK, Murphy JM, McBride WJ (2002) Cocaine is self-administered into the shell but not the core of the nucleus accumbens of Wistar rats. J Pharmacol Exp Ther 303:1216–1226

Rodriguez AL, Tarr JC, Zhou Y, Williams R, Gregory KJ, Bridges TM, Daniels JS, Niswender CM, Conn PJ, Lindsley CW, Stauffer SR (2010) Identification of a glycine sulfonamide based non-MPEP site positive allosteric potentiator (PAM) of mGlu5 probe reports from the NIH molecular libraries program, Bethesda (MD)

Rollema H, Chambers LK, Coe JW, Glowa J, Hurst RS, Lebel LA, Lu Y, Mansbach RS, Mather RJ, Rovetti CC, Sands SB, Schaeffer E, Schulz DW, Tingley FD 3rd, Williams KE (2007a) Pharmacological profile of the alpha4beta2 nicotinic acetylcholine receptor partial agonist varenicline, an effective smoking cessation aid. Neuropharmacology 52:985–994

Rollema H, Coe JW, Chambers LK, Hurst RS, Stahl SM, Williams KE (2007b) Rationale, pharmacology and clinical efficacy of partial agonists of alpha4beta2 nACh receptors for smoking cessation. Trends Pharmacol Sci 28:316–325

Rollema H, Shrikhande A, Ward KM, Tingley FD 3rd, Coe JW, O'Neill BT, Tseng E, Wang EQ, Mather RJ, Hurst RS, Williams KE, de Vries M, Cremers T, Bertrand S, Bertrand D (2010) Pre-clinical properties of the alpha4beta2 nicotinic acetylcholine receptor partial agonists varenicline, cytisine and dianicline translate to clinical efficacy for nicotine dependence. Br J Pharmacol 160:334–345

Rose JE (2006) Nicotine and nonnicotine factors in cigarette addiction. Psychopharmacology 184:274–285

Rose JE, Behm FM, Westman EC, Johnson M (2000) Dissociating nicotine and nonnicotine componenents of cigarette smoking. Pharmacol Biochem Behav 67:71–78

Salamone JD, Correa M (2012) The mysterious motivational functions of mesolimbic dopamine. Neuron 76:470–485

Salamone JD, Correa M (2013) Dopamine and food addiction: lexicon badly needed. Biol Psychiatry 73:e15–e24

Salamone JD, Correa M, Farrar A, Mingote SM (2007) Effort-related functions of nucleus accumbens dopamine and associated forebrain circuits. Psychopharmacology 191:461–482

Salamone JD, Correa M, Nunes EJ, Randall PA, Pardo M (2012) The behavioral pharmacology of effort-related choice behavior: dopamine, adenosine and beyond. J Exp Anal Behav 97:125–146

Saura J, Kettler R, Da Prada M, Richards JG (1992) Quantitative enzyme radioautography with 3H-Ro 41-1049 and 3H-Ro 19-6327 in vitro: localization and abundance of MAO-A and MAO-B in rat CNS, peripheral organs, and human brain. J Neurosci 12:1977–1999

Scherma M, Panlilio LV, Fadda P, Fattore L, Gamaleddin I, Le Foll B, Justinova Z, Mikics E, Haller J, Medalie J, Stroik J, Barnes C, Yasar S, Tanda G, Piomelli D, Fratta W, Goldberg SR (2008) Inhibition of anandamide hydrolysis by cyclohexyl carbamic acid 3′-carbamoyl-3-yl ester (URB597) reverses abuse-related behavioral and neurochemical effects of nicotine in rats. J Pharmacol Exp Ther 327:482–490

Scherma M, Justinova Z, Zanettini C, Panlilio LV, Mascia P, Fadda P, Fratta W, Makriyannis A, Vadivel SK, Gamaleddin I, Le Foll B, Goldberg SR (2012) The anandamide transport inhibitor AM404 reduces the rewarding effects of nicotine and nicotine-induced dopamine elevations in the nucleus accumbens shell in rats. Br J Pharmacol 165:2539–2548

Schilstrom B, Nomikos GG, Nisell M, Hertel P, Svensson TH (1998) N-methyl-D-aspartate receptor antagonism in the ventral tegmental area diminishes the systemic nicotine-induced dopamine release in the nucleus accumbens. Neuroscience 82:781–789

Schilstrom B, Rawal N, Mameli-Engvall M, Nomikos GG, Svensson TH (2003) Dual effects of nicotine on dopamine neurons mediated by different nicotinic receptor subtypes. Int J Neuropsychopharmacol 6:1–11

Schultz W (2004) Neural coding of basic reward terms of animal learning theory, game theory, microeconomics and behavioural ecology. Curr Opin Neurobiol 14:139–147

Schultz W (2007) Multiple dopamine functions at different time courses. Annu Rev Neurosci 30:259–288

Schultz W (2010) Dopamine signals for reward value and risk: basic and recent data. Behav Brain Functions 6:24

Schultz W (2011) Potential vulnerabilities of neuronal reward, risk, and decision mechanisms to addictive drugs. Neuron 69:603–617

Sellings LH, Clarke PB (2003) Segregation of amphetamine reward and locomotor stimulation between nucleus accumbens medial shell and core. J Neurosci 23:6295–6303

Sellings LH, McQuade LE, Clarke PB (2006) Evidence for multiple sites within rat ventral striatum mediating cocaine-conditioned place preference and locomotor activation. J Pharmacol Exp Ther 317:1178–1187

Sellings LH, Baharnouri G, McQuade LE, Clarke PB (2008) Rewarding and aversive effects of nicotine are segregated within the nucleus accumbens. Eur J Neurosci 28:342–352

Sesack SR, Grace AA (2010) Cortico-Basal Ganglia reward network: microcircuitry. Neuropsychopharmacology 35:27–47

Shoaib M, Stolerman IP (1999) Plasma nicotine and cotinine levels following intravenous nicotine self-administration in rats. Psychopharmacology 143:318–321

Shoaib M, Benwell ME, Akbar MT, Stolerman IP, Balfour DJ (1994) Behavioural and neurochemical adaptations to nicotine in rats: influence of NMDA antagonists. Br J Pharmacol 111:1073–1080

Shoaib M, Schindler CW, Goldberg SR (1997) Nicotine self-administration in rats: strain and nicotine pre-exposure effects on acquisition. Psychopharmacology 129:35–43

Shoaib M, Sidhpura N, Shafait S (2003) Investigating the actions of bupropion on dependence-related effects of nicotine in rats. Psychopharmacology 165:405–412

Singer G, Wallace M (1984) Effects of 6-OHDA lesions in the nucleus accumbens on the acquisition of self injection of heroin under schedule and non schedule conditions in rats. Pharmacol Biochem Behav 20:807–809

Singer G, Wallace M, Hall R (1982) Effects of dopaminergic nucleus accumbens lesions on the acquisition of schedule induced self injection of nicotine in the rat. Pharmacol Biochem Behav 17:579–581

Spiller K, Xi ZX, Li X, Ashby CR Jr, Callahan PM, Tehim A, Gardner EL (2009) Varenicline attenuates nicotine-enhanced brain-stimulation reward by activation of alpha4beta2 nicotinic receptors in rats. Neuropharmacology 57:60–66

Stead LF, Lancaster T (2007) Interventions to reduce harm from continued tobacco use. Cochrane Database Syst Rev: CD005231

Stead LF, Perera R, Bullen C, Mant D, Hartmann-Boyce J, Cahill K, Lancaster T (2012) Nicotine replacement therapy for smoking cessation. Cochrane Database Syst Rev 11:CD000146

Stefanik MT, Kupchik YM, Brown RM, Kalivas PW (2013) Optogenetic evidence that pallidal projections, not nigral projections, from the nucleus accumbens core are necessary for reinstating cocaine seeking. J Neurosci 33:13654–13662

Stuber GD, Roitman MF, Phillips PE, Carelli RM, Wightman RM (2005) Rapid dopamine signaling in the nucleus accumbens during contingent and noncontingent cocaine administration. Neuropsychopharmacology 30:853–863

Suemaru K, Gomita Y, Furuno K, Araki Y (1993) Chronic nicotine treatment potentiates behavioral responses to dopaminergic drugs in rats. Pharmacol Biochem Behav 46:135–139

Sutherland G, Russell MA, Stapleton J, Feyerabend C, Ferno O (1992) Nasal nicotine spray: a rapid nicotine delivery system. Psychopharmacology 108:512–518

Taepavarapruk P, Floresco SB, Phillips AG (2000) Hyperlocomotion and increased dopamine efflux in the rat nucleus accumbens evoked by electrical stimulation of the ventral subiculum: role of ionotropic glutamate and dopamine D1 receptors. Psychopharmacology 151:242–251

Tessari M, Pilla M, Andreoli M, Hutcheson DM, Heidbreder CA (2004) Antagonism at metabotropic glutamate 5 receptors inhibits nicotine- and cocaine-taking behaviours and prevents nicotine-triggered relapse to nicotine-seeking. Eur J Pharmacol 499:121–133

Tronci V, Balfour DJ (2011) The effects of the mGluR5 receptor antagonist 6-methyl-2-(phenylethynyl)-pyridine (MPEP) on the stimulation of dopamine release evoked by nicotine in the rat brain. Behav Brain Res 219:354–357

Tronci V, Vronskaya S, Montgomery N, Mura D, Balfour DJ (2010) The effects of the mGluR5 receptor antagonist 6-methyl-2-(phenylethynyl)-pyridine (MPEP) on behavioural responses to nicotine. Psychopharmacology 211:33–42

Van Gucht D, Van den Bergh O, Beckers T, Vansteenwegen D (2010) Smoking behavior in context: where and when do people smoke? J Behav Ther Exp Psychiatry 41:172–177

Vanderschuren LJ, Kalivas PW (2000) Alterations in dopam mergic and glutamatergic transmission in the induction and expression of behavioural sensitization: a critical review of preclinical studies. Psychopharmacology 151:99–120

Vansickel AR, Eissenberg T (2013) Electronic cigarettes: effective nicotine delivery after acute administration. Nicotine Tob Res 15:267–270

Villegier AS, Lotfipour S, McQuown SC, Belluzzi JD, Leslie FM (2007) Tranylcypromine enhancement of nicotine self-administration. Neuropharmacology 52:1415–1425

Volkow ND, Wang GJ, Fowler JS, Tomasi D, Telang F (2011) Addiction: beyond dopamine reward circuitry. Proc Natl Acad Sci USA 108:15037–15042

Warner C, Shoaib M (2005) How does bupropion work as a smoking cessation aid? Addict Biol 10:219–231

Watkins SS, Stinus L, Koob GF, Markou A (2000) Reward and somatic changes during precipitated nicotine withdrawal in rats: centrally and peripherally mediated effects. J Pharmacol Exp Ther 292:1053–1064

Wickham R, Solecki W, Rathbun L, McIntosh JM, Addy NA (2013) Ventral tegmental area alpha6beta2 nicotinic acetylcholine receptors modulate phasic dopamine release in the nucleus accumbens core. Psychopharmacology 229:73–82

Willuhn I, Wanat MJ, Clark JJ, Phillips PE (2010) Dopamine signaling in the nucleus accumbens of animals self-administering drugs of abuse. Curr Top Behav Neurosci 3:29–71

Wing VC, Shoaib M (2008) Contextual stimuli modulate extinction and reinstatement in rodents self-administering intravenous nicotine. Psychopharmacology 200:357–365

Wise RA (1987) The role of reward pathways in the development of drug dependence. Pharmacol Ther 35:227–263

Wise RA (1988a) The neurobiology of craving: implications for the understanding and treatment of addiction. J Abnorm Psychol 97:118–132

Wise RA (1988b) Psychomotor stimulant properties of addictive drugs. Ann NY Acad Sci 537:228–234

Wise RA (1996) Neurobiology of addiction. Curr Opin Neurobiol 6:243–251

Wise RA, Bozarth MA (1985) Brain mechanisms of drug reward and euphoria. Psychiatr Med 3:445–460

Wise RA, Bozarth MA (1987) A psychomotor stimulant theory of addiction. Psychol Rev 94:469–492

Wouda JA, Riga D, De Vries W, Stegeman M, van Mourik Y, Schetters D, Schoffelmeer AN, Pattij T, De Vries TJ (2011) Varenicline attenuates cue-induced relapse to alcohol, but not nicotine seeking, while reducing inhibitory response control. Psychopharmacology 216:267–277

Zahm DS, Brog JS (1992) On the significance of subterritories in the “accumbens” part of the rat ventral striatum. Neuroscience 50:751–767

Zhang T, Zhang L, Liang Y, Siapas AG, Zhou FM, Dani JA (2009) Dopamine signaling differences in the nucleus accumbens and dorsal striatum exploited by nicotine. J Neurosci 29:4035–4043

Zhang L, Dong Y, Doyon WM, Dani JA (2012) Withdrawal from chronic nicotine exposure alters dopamine signaling dynamics in the nucleus accumbens. Biol Psychiatry 71:184–191

Nicotine Withdrawal

Ian McLaughlin, John A. Dani and Mariella De Biasi

Abstract An aversive abstinence syndrome manifests 4–24 h following cessation of chronic use of nicotine-containing products. Symptoms peak on approximately the 3rd day and taper off over the course of the following 3–4 weeks. While the severity of withdrawal symptoms is largely determined by how nicotine is consumed, certain short nucleotide polymorphisms (SNPs) have been shown to predispose individuals to consume larger amounts of nicotine more frequently—as well as to more severe symptoms of withdrawal when trying to quit. Additionally, rodent behavioral models and transgenic mouse models have revealed that specific nicotinic acetylcholine receptor (nAChR) subunits, cellular components, and neuronal circuits are critical to the expression of withdrawal symptoms. Consequently, by continuing to map neuronal circuits and nAChR subpopulations that underlie the nicotine withdrawal syndrome—and by continuing to enumerate genes that predispose carriers to nicotine addiction and exacerbated withdrawal symptoms—it will be possible to pursue personalized therapeutics that more effectively treat nicotine addiction.

Keywords Nicotine withdrawal · SNP · Behavior · Medial habenula · Interpeduncular nucleus · Nicotinic subunits

---

I. McLaughlin · M. De Biasi (✉)
Department of Psychiatry, Perelman School of Medicine, University of Pennsylvania, Philadelphia, PA 19104, USA
e-mail: marielde@upenn.edu

J.A. Dani · M. De Biasi
Department of Neuroscience, Perelman School of Medicine, University of Pennsylvania, Philadelphia, PA 19104, USA

J.A. Dani · M. De Biasi
Mahoney Institute for Neurosciences, Perelman School of Medicine, University of Pennsylvania, Philadelphia, PA 19104, USA

I. McLaughlin
Neuroscience Graduate Group, Perelman School of Medicine, University of Pennsylvania, Philadelphia, PA 19104, USA

D.J.K. Balfour and M.R. Munafò (eds.), *The Neuropharmacology of Nicotine Dependence*, Current Topics in Behavioral Neurosciences 24,
DOI 10.1007/978-3-319-13482-6_4

Contents

1 Withdrawal Syndrome in Humans

1.1 Symptoms

The Diagnostic and Statistical Manual of Mental Disorders (DSM-5) reports 7 primary symptoms associated with nicotine withdrawal: irritability/anger/frustration, anxiety, depressed mood, difficulty concentrating, increased appetite, insomnia, and restlessness (American Psychiatric Association 2013). The syndrome might also include constipation, dizziness, nightmares, nausea, and sore throat. For practical purposes, nicotine withdrawal symptoms are classified as affective, somatic, and cognitive. Affective symptoms include anxiety, anhedonia, depression, dysphoria, hyperalgesia, and irritability. Somatic manifestations include tremors, bradycardia, gastrointestinal discomfort, and increased appetite. Cognitive symptoms manifest as difficulty concentrating and impaired memory (Heishman et al. 2010). This constellation of symptoms reflects the brain-wide influence of cholinergic transmission. Studies characterizing the participation of specific nAChRs in signaling during particular manifestations of withdrawal are helping to reveal their underlying mechanisms (Paolini and De Biasi 2011).

1.2 Genetic Influences

Nicotine addiction is influenced by both genetic and environmental factors. Depending on the parameters used to define dependence, and the population considered, heritability contributes 50–75 % of the risk for dependence (Dokal et al. 1989;

Hall et al. 2002; Lessov et al. 2004; Furberg et al. 2010; Nugent et al. 2014; also see chapters entitled Genetics of Smoking Behaviour and Pharmacogenetics of Nicotine and Associated Smoking Behaviors; volume 23). Risk factors for nicotine dependence can be identified using genetic methods such as linkage and candidate gene analyses, as well as genome-wide association studies (GWAS). Linkage analysis assesses the presence of a phenotype in large, high-risk families to map the location of disease-causing loci in relation to a known genetic marker. Candidate gene analysis assesses the association between a particular gene allele (or alleles) potentially involved in the disease and the disease itself. GWAS, in contrast to candidate gene approaches, do not limit analyses to relationships between specific genes and a phenotype and aim to identify loci for novel susceptibility genes. An example of how such genetic approaches can be successfully utilized in nicotine research is the gene cluster on chromosome 15q25 that encodes the α5, α3, and β4 nAChR subunits. Gene variants within the cluster have been repeatedly shown to affect nicotine dependence and smoking quantity (Saccone et al. 2009). A non-synonymous nucleotide polymorphism (SNP), rs16969968, which substitutes an aspartic acid for an asparagine (D398N) in the *CHRNA5* region of the cluster, has been associated with reduced receptivity to nAChR agonists in vitro, reduced neuronal calcium permeability, and more extensive nAChR desensitization (Jackson et al. 2010). Individuals who are homozygous for this SNP are more likely to progress to heavy smoking and nicotine dependence (Hartz et al. 2012), and it has been suggested that the reduction in agonist responsivity and increased desensitization may contribute to these increased propensities (Jackson et al. 2010). Apart from the missense rs16969968 variant, there are other SNPs within the *CHRNA5-CHRNA3-CHRNB4* cluster that are relevant to nicotine addiction and withdrawal, and further biological characterizations are needed to establish the functional consequences of those mutations. Other nAChR genes have also been shown to influence smoking quantity and nicotine dependence, such as common variants in the chromosome 8p11 region that contains the genes encoding the α6 and β3 nAChR subunits (Thorgeirsson et al. 2010). Overall, linkage analyses have highlighted 13 regions on 11 chromosomes that include genes with potential influences on nicotine dependence (Leeb and Tamse 1985; Li 2008). Genes involved in nicotine metabolism are also likely important to the nature of nicotine withdrawal. Cytochrome P450 2A6 (CYP2A6) is the enzyme mainly responsible for the conversion of nicotine to cotinine, which typically accounts for 70–80 % of nicotine metabolism (Hecht et al. 2000). In subjects of European descent, GWAS meta-analyses identified SNPs in the region of *CYP2A6* associated with the number of cigarettes smoked per day, as well as other smoking behavior phenotypes (Thorgeirsson et al. 2010).

Genetic factors may also account for 29–53 % of the variance in withdrawal symptoms and approximately 50 % of the variance in quitting success (Xian et al. 2003, 2005; Pergadia et al. 2006). In a linkage analysis study, a sample of Australian and Finnish smokers was queried about withdrawal symptoms within the context of a smoking cessation attempt that they recalled well. The study revealed a linkage signal that meets genome-wide significance on chromosome 11p15 in the Finnish families (Pergadia et al. 2009). Four strong candidate genes lie within or near the peak area on chromosome 11: dopamine receptor 4 (*DRD4*), TPH tryptophan hydroxylase 1 (*TPH*), tyrosine hydroxylase (*TH*), and nAChR subunit 10 (*CHRNA10*). A second region identified by linkage on chromosome 11q23 includes hydroxytryptamine receptor 3A (*HTR3A*), hydroxytryptamine receptor 3B (*HTR3B*), dopamine D2 receptor gene (*DRD2*), ankyrin repeat and kinase domain containing 1(*ANKK1*), and ionotropic kainate glutamate receptor 4 gene (*GRIK4*). An earlier study reported that *DRD2* TaqI-B polymorphisms influence abstinence and withdrawal symptoms (Robinson et al. 2007). Smokers carrying the *DRD2* TaqI-B1 risk allele (B1/B1 or B1/B2) reported significantly more symptoms of daily smoking withdrawal compared to smokers homozygous for the TaqI-B2 allele (B2/B2). In addition, while withdrawal symptoms—measured 14 days pre-quit and 42 days post-quit—decreased significantly in Taql-B2 homozygous over time, smokers with the TaqI-B1 allele reported little improvement in self-reported withdrawal symptoms.

Several other pharmacogenetic studies, including a couple of genome-wide association study (GWAS) analyses, have examined genetic influences on smoking cessation and response to therapy (Leeb and Tamse 1985; Uhl et al. 2007; Furberg et al. 2010; Gold and Lerman 2012; King et al. 2012; Bloom et al. 2013; Chen et al. 2014). SNPs in *CHRNB2* and the *CHRNA5-CHRNA3-CHRNB4* cluster seem to influence smoking cessation, although the reported effects are not always reproducible (Conti et al. 2008; Baker et al. 2009; Perkins et al. 2009; Gold and Lerman 2012; Hartz et al. 2012). The influence of *CHRNA5-CHRNA3-CHRNB4* haplotypes on tolerance, craving, and loss of control seems greatest in individuals who began smoking early in life, suggesting that the risk associated with those genes is greatest with early tobacco exposure (Baker et al. 2009). The rs8192475 variant in *CHRNA3* was shown to predict withdrawal symptoms and craving after quitting (Sarginson et al. 2011). The major G allele was associated with stronger but relatively short-lived symptoms, while the minor A allele resulted in a relatively constant level of symptoms and craving during the acute phase of withdrawal. The rs8192475 α3 SNP leads to an R37H change in the amino acid sequence that increases agonist sensitivity of heterologously expressed α3β4 nAChRs (Haller et al. 2012). Rare missense variants at conserved residues in *CHRNB4* (T375I and T91I) are associated with reduced numbers of cigarettes per day and fewer signs of withdrawal (Haller et al. 2012). Similar to the α3 SNP rs8192475, expression of the two β4 mutations in HEK 293 cells leads to larger currents in response to ACh stimulation, which has been suggested to increase the aversive properties of nicotine.

Similar to what was found for phenotypes related to nicotine dependence, there are also associations of CYP2A6 enzyme activity and nicotine metabolic rates with smoking cessation and smoking cessation strategies (Lee et al. 2007; Chen et al. 2014). Individuals carrying the null activity allele, *CYP2A6*2*, are twice as likely to quit smoking as subjects without that allele (Gu et al. 2000). Conversely, smokers with high-activity alleles (*CYP2A6*1/*1B*) experience more severe withdrawal symptoms when trying to quit (Kubota et al. 2006).

Currently, the literature suggests that multiple genetic loci influence the severity of withdrawal symptoms and the ability to abstain from smoking. Such polygenic contributions overlap, at least in part, with those associated with vulnerability to nicotine dependence. Because the majority of published studies was conducted on relatively small samples, examined different cessation strategies, and did not always address withdrawal symptoms directly, additional work is needed to yield more robust and reproducible associations between genes and withdrawal phenotypes. Once putative genes are identified, it will be critical to assess the potential for genetic variations to alter biological function and, consequently, nicotine-related behaviors. Preclinical studies that address both behavior and molecular mechanisms can help to explain how the changes in DNA sequence are associated with the behavioral phenotypes observed in smokers. Rodents, and particularly mice, offer unique opportunities to explore and characterize the relationships between gene function and nicotine withdrawal. Given the difficulty of effectively characterizing gene products and circuits responsible for the symptomatology of nicotine withdrawal in humans, genetic manipulations can be carried out in mice to directly address the circuits and molecular mechanisms involved.

2 Withdrawal Syndrome in Mice

Many symptoms of nicotine withdrawal described in humans can be recapitulated in rodent models of addiction and withdrawal under experimental conditions (see chapter entitled The Role of Mesoaccumbens Dopamine in Nicotine Dependence; this volume). After at least a week of chronic nicotine administration in mice, cessation of nicotine exposure or administration of a nAChR antagonist induces affective, somatic, and cognitive signs of withdrawal (Malin et al. 1994; De Biasi and Salas 2008). The withdrawal syndrome exhibited can be alleviated by nicotine administration—akin to nicotine replacement strategies such as nicotine patches or gum—or by administration of pharmacological agents used as cessation aids in humans, such as bupropion or varenicline (Rennard and Daughton 2014). Because mice express neuroadaptations that result in the nicotine withdrawal syndrome, and because nAChR subunit knockout mouse lines are readily available, researchers have been able to study the functional contributions of particular nAChR subunits to specific symptoms of withdrawal.

2.1 Behavioral Manifestations and Testing Paradigms

The nicotine withdrawal syndrome can be studied in mice by observing the frequencies of certain stereotypies, or by evaluating changes in behavior during withdrawal, relative to baseline behaviors exhibited by control mice naïve to nicotine (see chapter entitled Behavioral Mechanisms Underlying Nicotine Reinforcement; this volume). As in humans, the behavioral manifestations of nicotine withdrawal in mice can be categorized as somatic, affective, and cognitive.

Somatic symptoms in mice include excessive grooming, chewing, tremors, "wet-dog" shakes, yawns, and teeth chattering (Paolini and De Biasi 2011). The occurrence of these symptoms tends to increase according to the severity of withdrawal. Affective symptoms are somewhat subtler in mice than somatic symptoms; their evaluation requires a series of behavioral assays to reveal changes in affect that are associated with withdrawal. Just as humans undergoing nicotine withdrawal experience anxiety, anhedonia, depression, and hyperalgesia, behavioral paradigms applied to mice expose analogous affective states. A challenge researchers face when evaluating rodent affect is that no single paradigm effectively isolates any one particular affective state. Each measure likely evokes multiple affective states, unintentionally recruiting off-target neuronal circuits that are outside the scope of a given study. However, several behavioral paradigms applied as a set can help reveal relative differences in affect following experimental treatments. The emergence of anxiety-like symptoms following nicotine abstinence can be measured in the open field test (OFT) and the elevated plus maze (EPM), two of the most commonly employed paradigms in studies evaluating emotional states of rodents (Crawley et al. 1997). The OFT is conducted by placing a mouse or rat in a square arena and evaluating the ratio of time rodents spend in the center of the environment to time spent along its perimeter. As rodents experience increased anxiety, they tend to avoid the center and spend more time along the walls of the environment, a behavior termed thigmotaxis. As rodents undergo nicotine withdrawal, they tend to exhibit increased levels of thigmotaxis. The OFT also enables researchers to evaluate rearing behavior, overall locomotion, freezing, and changes in defecation during withdrawal. As a further source of experimental control, levels of ambient illumination can be adjusted to modulate baseline levels of anxiety. The EPM consists of two enclosed arms with walls running along their periphery, and two open arms with no walls that present an environment resembling elevated cliffs (Crawley et al. 1997). The arms are arranged in the shape of a plus sign, and anxious mice tend to make fewer entries into the open arms and spend less time in open arms. Rodents undergoing nicotine withdrawal exhibit both of these anxiety-associated behaviors, and anxiolytic drugs have been observed to increase open-arm entries (Crawley et al. 1997).

Conditioned place aversion (CPA) can be used to measure the dysphoric manifestations associated with withdrawal. The paradigm involves repeated pairing of distinct environmental cues with negative stimuli (i.e., withdrawal), such that when given a choice, mice avoid the withdrawal-paired cues relative to neutral cues. The time mice spend avoiding environments paired with withdrawal serves as an indicator of the severity of aversion (Kenny and Markou 2001). The emergence of depression-like symptoms during withdrawal can be assessed with the forced swim test (FST), which assesses learned helplessness by monitoring passive coping strategies such as immobility (Cryan et al. 2002; Thanos et al. 2013). During withdrawal, immobility is increased, indicating a depression-like state.

Intracranial self-stimulation (ICSS) is used to investigate the "reward" circuitry and can be useful to evaluate anhedonia during withdrawal. Electrodes are implanted into the rodent brain targeted to the medial forebrain bundle, which includes the mesolimbic DA pathway associated with hedonia or reward. The rodent is first trained to perform the operant task of self-stimulation and then is allowed to self-administer small electrical stimulations to the targeted pathway. As rodents perform this operant conditioning, the stimulus intensity is experimentally adjusted to determine the baseline stimulation threshold at which self-stimulation is consistently achieved and retained. Following the establishment of this threshold, researchers can explore the effects various stimuli have on brain reward and hedonic signaling. A lowered threshold resulting from a treatment is suggested to represent increased reward signaling along the pathway, while a heightened threshold is interpreted as indicating a state of greater anhedonia (O'Dell and Khroyan 2009). Animals experiencing nicotine withdrawal display elevated thresholds for ICSS.

The effects of withdrawal on cognition, especially hippocampal-dependent learning and memory, can be examined with contextual fear conditioning (FC). Subjects must learn contextual information and form an association between the context and an aversive electric shock (Fanselow and Poulos 2005; Sigurdsson et al. 2007). Acute nicotine administration enhances contextual FC, while nicotine withdrawal impairs it (Davis and Gould 2008).

The five-choice serial reaction time task (5-CSRTT), also called operant signal detection, is a behavioral test used to characterize the effects of treatments on sustained and divided attention (Robbins 2002). The task consists of a maze that presents five holes, each of which may be paired with a reward. A brief flash of light inside one of the holes indicates the one that contains a reward, and rodents must poke the correct hole to receive the reward. This task quantifies the capability of rodents to maintain spatial attention that has been split among five locations over the course of many trials by dividing the proportion of correct nose-pokes (pokes into the lit hole) by total nose-pokes. While nicotine administration has been observed to generate cognitive enhancements, rats withdrawing from nicotine have been observed to exhibit attention deficits in the 5-CSRTT paradigm (Shoaib and Bizarro 2005), reflecting some of the symptoms that human smokers report while undergoing withdrawal.

While these rodent behaviors may reflect symptoms of the human withdrawal syndrome, the neuronal circuits recruited during experimental paradigms remain largely unknown. Differences in the behavioral outputs between animals undergoing withdrawal and control subjects can help reveal the circuits nicotine withdrawal impinges upon, as each behavioral test offers unique environmental stimuli and engages specific neuronal circuits (Wahlsten 2010). The integration of mice genetically modified to carry nAChR null [knockout (KO)] or SNP-related mutations into behavioral research has introduced a finer level of resolution to the characterization of the neuronal basis of withdrawal. By examining how eliminating signaling contributions from specific receptor subunits affects behavioral outcomes, researchers can determine which receptor subunits are necessary for the expression of certain nicotine withdrawal symptoms. Additionally, the genetic differences that distinguish each inbred strain of mice have important consequences for behavior, as each strain is characterized by a unique repertoire of consistent behavioral traits (Crawley 1996, 2008; Krackow et al. 2010; Matsuo et al. 2010). As these differences can translate to increased or decreased detectability of behavioral changes induced by withdrawal, it is prudent to establish which strain is most sensitive to the hypothesized behavioral changes in question (Bailey et al. 2006; Crawley 2008; Lalonde and Strazielle 2008; Wahlsten 2010).

3 Receptors and nAChR Subunits Underlying Withdrawal Symptoms

3.1 *Insight from nAChR Subunit KO Mice*

Ultimately, the development of enhanced therapeutics that effectively promote smoking cessation with minimal side effects will result from the identification of particular receptor subunit compositions that are necessary for the expression of withdrawal symptoms. Mouse models carrying nAChR null mutations are helping to identify the nAChR subtypes responsible for the various symptoms of nicotine withdrawal.

In rodents, α2 mRNA levels are highest in the interpeduncular nucleus (IPN), with more restricted expression in scattered neurons within the amygdala, hippocampus, cortex, retina, spinal cord, and cerebellum (Lotfipour et al. 2013). Analyses of withdrawal symptoms in α2 KO mice have shown that the physical manifestations of nicotine abstinence are context-dependent, as α2 KO mice exhibit more somatic symptoms of withdrawal in novel environments (Lotfipour et al. 2013), but display no somatic signs of withdrawal in habituated environments (i.e., the home cage) (Salas et al. 2009). Additionally, the lack of α2 is sufficient to abolish somatic symptoms of precipitated nicotine withdrawal (Salas et al. 2009). In male mice, *Chrna2* deletion also produces nicotine withdrawal-induced deficits of cued FC (Lotfipour et al. 2013). α2 KO mice also self-administer higher doses of nicotine than WT controls (Lotfipour et al. 2013). It should be noted that in a study of European Americans and African-Americans, *CHRNA2* showed a strong association with the Fagerström Test for Nicotine Dependence (FTND) after correction for multiple testing (Wang et al. 2014). The rs2472553 *SNP*, which seems to have the strongest association with nicotine dependence, encodes a functional variant in the signal peptide, which leads to a threonine-to-isoleucine amino acid substitution at residue 22. In oocytes, the T22I mutation changes the sensitivity to nicotine of α2β4-containing nAChRs (Dash et al. 2014).

The α3 nAChR *subunit* is encoded by *CHRNA3*, one of the genes in the chromosome 15q25 cluster that has shown the most robust association with smoking behavior and nicotine dependence. α3 forms functional nAChR complexes with α5 and β4, the other two subunits encoded by *CHRNA5-CHRNA3-CHRNB4*. These subunits are densely expressed in the medial habenula (MHb) and IPN (Grady et al. 2009). Due to developmental abnormalities—including bladder enlargement and infection, urinary stones, and difficult urination—α3 KO mice survive birth, but exhibit severely impaired growth and perinatal mortality (Xu et al. 1999).

This phenotype has rendered evaluation of the subunit's potential involvement in withdrawal symptomatology impractical in α3 null mice. However, pharmacological data suggest that receptors comprising the α3 and β4 nAChR subunits influence both nicotine reward and somatic manifestations of withdrawal (Jackson et al. 2013). AuIB, an α-conotoxin peptide that potently blocks α3β4 nAChRs (Luo et al. 1998), dose-dependently inhibits nicotine elicited reward as measured in the conditioned place preference paradigm. The α-conotoxin also reduces somatic signs of withdrawal and withdrawal-induced hyperalgesia, while it has no effect on the aversive motivational component of withdrawal as measured in the CPA paradigm (Jackson et al. 2013).

α5-containing nAChRs are expressed in various brain areas implicated in the key effects of nicotine, including ventral tegmental area (VTA), MHb, IPN, hippocampus, and cortex (Salas et al. 2003a). α5 KO mice do not display physical symptoms associated with both spontaneous and mecamylamine-precipitated nicotine withdrawal (Salas et al. 2009) nor withdrawal-induced hyperalgesia (Jackson et al. 2008). The MHb/IPN pathway plays a crucial role in mediating the physical manifestations of nicotine withdrawal and is the likely location of the effects of α5-containing nAChR on this phenotype (Salas et al. 2009). Acute nicotine application to MHb slices enhances the intrinsic excitability of MHb neurons (Dao et al. 2014). This dynamic depends on the presence of α5-containing nAChRs within the MHb and the release of neurokinins (Dao et al. 2014), as such enhancement of excitability was prevented by bath application of neurokinin 1 (NK1) or NK3 receptor antagonists. In addition, infusion of the same NK receptor antagonists into the MHb of mice chronically treated with nicotine precipitated somatic signs of nicotine withdrawal (Dao et al. 2014). Microinjection of neurokinin receptor antagonists into adjacent anatomical structures, including the lateral habenula (LHb), failed to elicit behavior resembling withdrawal. Similarly, microinjection of the NK1 and/or NK3 antagonists into the MHb of nicotine-naïve mice failed to generate somatic symptoms of withdrawal. It was concluded that interactions between cholinergic and neurokininergic systems contribute to the emergence of nicotine withdrawal symptoms (Dao et al. 2014). The α5 null mutation does not appear to influence affective symptoms such as withdrawal-induced CPA (Jackson et al. 2008). The same study reported that α5 KO mice do not display increased anxiety levels in the EPM during withdrawal (Jackson et al. 2008). However, interpretation of those data is not straightforward, as *Chrna5* deletion reduces anxiety in the EPM in basal conditions independent of nicotine treatment (Gangitano et al. 2009). It has also been shown that α5-containing nAChRs do not regulate the reward-inhibiting effects induced by nicotine withdrawal in the ICSS paradigm (Fowler et al. 2013), suggesting that the receptors do not influence anhedonia.

The α6 subunit also appears to play a role in the nicotine withdrawal syndrome. DA release in the NAcc following nicotine administration is regulated in part by α6-containing (α6*) nAChRs on DAergic terminals in the dorsal and ventral striatum (Exley et al. 2008) and DAergic somata in the VTA (Grady et al. 2007; Zhao-Shea et al. 2011). Intracerebral infusion of a selective antagonist of α6β2* nAChR blocks CPA and withdrawal-precipitated anxiety-like behavior in the EPM (Jackson et al. 2009). There was no influence on the somatic symptoms of withdrawal, suggesting a selective role for α6 nAChR subunits in the affective manifestations of withdrawal (Jackson et al. 2009).

Unlike most other subunits, α7 nAChR subunits are capable of forming homomeric receptors that are broadly distributed in the brain. The hyperalgesia symptoms that emerge during mecamylamine-precipitated withdrawal are reduced in α7 KO mice (Grabus et al. 2005). In contrast to wild-type mice, α7 null mice failed to exhibit elevated ICSS thresholds during nicotine withdrawal between 3 and 6 h after their last nicotine exposure (Stoker et al. 2012). When the anhedonic affective state was evaluated between 8 and 100 h after nicotine withdrawal, ICSS thresholds were equally elevated in α7 WT and null littermates, indicating that a lack of α7 nAChR subunits delays, rather than abolishes, withdrawal symptoms. As for the physical signs of abstinence, α7 null mice exhibit significantly reduced withdrawal symptoms immediately after precipitation of withdrawal by mecamylamine injection (Salas et al. 2007). However, their physical signs are indistinguishable from those of WT mice when measured at later times, up to 48 h after withdrawal (Jackson et al. 2008; Stoker et al. 2012).

β2 KO mice exhibit levels of somatic signs of withdrawal and abstinence-induced hyperalgesia comparable to those of WT mice (Salas et al. 2004). However, the mutant mice do not exhibit anxiety-related behaviors normally associated with withdrawal from chronic nicotine exposure (Jackson et al. 2008). Overall, these results suggest participation of β2 nAChR subunits in the signaling responsible for affective, but not somatic, symptoms of nicotine withdrawal. α6β2* nAChRs are expressed in the VTA and ventral striatum and are associated with reward and addiction, but they are not expressed peripherally. Therefore α6β2* nAChRs may represent viable targets for the treatment of affective symptoms experienced during nicotine withdrawal. Indeed, a non-nicotine pharmaceutical currently FDA approved for the treatment of smoking cessation, varenicline, acts as a partial agonist at α6β2* nAChRs (Grady et al. 2010; Bordia et al. 2012).

While sequence variants associated with smoking behavior lie within the regions that harbor the *CHRNB3-CHRNA6* genes on chromosome 8p11 (Thorgeirsson et al. 2010), no preclinical data are available on the effects of the β3 null mutation on nicotine-related behavior. However, it has been determined that β3 KO mice exhibit lower baseline levels of anxiety-related behavior in three different paradigms (Booker et al. 2007). As reported for the α5 KO mice (Gangitano et al. 2009), β3 KO mice have altered hypothalamic–pituitary–adrenal axis responses (Booker et al. 2007). Changes were also reported for locomotor activity and prepulse inhibition of acoustic startle, behaviors that are controlled, at least in part, by nigrostriatal and mesolimbic dopaminergic activity (Cui et al. 2003). As β3 mRNA is detected in the substantia nigra, VTA, and medial habenula (Cui et al. 2003), it is tempting to attribute the anxiolytic phenotype to MHb mechanisms and the locomotor phenotype to nigrostriatal mechanisms.

β4 KO mice exhibit no somatic signs of nicotine withdrawal or hyperalgesia following mecamylamine-induced withdrawal (Salas et al. 2004). Similar results were found for β4 KO mice undergoing spontaneous withdrawal from chronic nicotine administration (Stoker et al. 2012). In addition, β4 KO mice do not display anhedonia-like symptoms during withdrawal, as identified by unchanged intracranial self-stimulation thresholds (Stoker et al. 2012). The reported phenotypes likely reflect an involvement of different subtypes of β4* nAChRs. α3β4* nAChRs are the most likely contributors to physical components of withdrawal given the high levels of expression of α3 and β4 mRNA in the MHb, and the fact that the α3β4-selective α-conotoxin AuIB blocks the emergence of somatic signs of withdrawal (Jackson et al. 2013). Because AuIB does not interfere with the affective symptomatology of withdrawal in the EPM and CPA paradigms, α6β4* nAChRs might be involved. Indeed, a transgenic mouse model that overexpresses β4 nAChR subunits exhibits altered nicotine consumption and CPA (Frahm et al. 2011), and there is a documented role for α6* nAChRs in withdrawal-induced CPA (Jackson et al. 2009).

4 Molecular Mechanisms Involved in the Nicotine Withdrawal Syndrome

4.1 nAChR Upregulation

Long-term exposure to nicotine leads to an increase, or upregulation, of nicotine-binding sites in the brain of smokers (Benwell et al. 1988) and rodents subjected to repeated nicotine administrations (Marks et al. 1983). The increased pool of nAChRs arising from chronic nicotine exposure may drive symptoms associated with the nicotine withdrawal syndrome (Turner et al. 2011; Gould et al. 2012) and might impact the ability to maintain abstinence in the clinical population (Staley et al. 2006). Using a radioligand with specificity for β2-containing nAChRs, which enabled the use of single-photon emission computed tomography (SPECT) (Staley et al. 2006), it was found that chronic smokers have more cortical, striatal, and cerebellar β2* nAChRs than non-smokers (Staley et al. 2006). Furthermore, β2* nAChRs in the anterior cingulate and frontal cortex were significantly correlated with the number of days following cessation of smoking. This was interpreted as a progressive increase of the number of β2* nAChRs in proportion to the number of consecutively abstinent days. In addition, a significant negative correlation was observed between β2* nAChR availability in the post-central gyrus or somatosensory cortex and the urge to relieve withdrawal symptoms by smoking (Staley et al. 2006). Similarly, in rodents, nAChR upregulation in the dorsal hippocampus was associated with withdrawal-related deficits in hippocampal learning (Gould et al. 2012; Portugal et al. 2012).

A variety of cellular processes influence receptor upregulation (Govind et al. 2009; Rezvani et al. 2009, 2010; Henderson et al. 2014). Receptors containing the β2 subunits are particularly sensitive to nicotine-induced upregulation. If β4 replaces β2 subunits in either α3* or α4* nAChRs, receptor upregulation is significantly reduced (Wang et al. 1998; Sallette et al. 2004). As replacement of β2 by β4 is sufficient to increase nAChR levels at the plasma membrane in the absence of nicotine, the proposed "chaperoning" function of nicotine might not be as effective (Srinivasan et al. 2011; Henderson et al. 2014). The presence of an accessory nAChR subunit in the receptor complex can also influence nicotine-induced upregulation. For example, the presence of α5 renders α4β2* nAChRs insensitive to nicotine-induced upregulation (Mao et al. 2008). Conversely, co-expression of β3 increases α6β2 and α6β4 receptor levels and enhances nicotine-induced upregulation of α6β2β3 receptors compared to α6β2 receptors (Tumkosit et al. 2006). The effects of β3 on receptor trafficking and upregulation may help explain why in the striatum α6-containing receptors without β3 are downregulated by nicotine, while those containing β3 are unaffected (Perry et al. 2007).

4.2 Desensitization of nAChRs

While some combinations of nAChR subunits render receptor complexes more prone to upregulation than others, desensitization is a prominent mechanism that contributes to this upregulation (Fenster et al. 1999). As nicotine has a half-life in humans of 2 h or more, nicotine accumulates during a day of regular smoking to reach steady-state plasma concentrations typically ranging between 10 and 50 ng/mL (Graham et al. 2007). This long-lasting level of nicotine ensures that high-affinity nAChRs recurrently bind and unbind nicotine throughout the day (Picciotto et al. 2008). Consequently, receptors undergo transitions between different conformations in response to ligand binding and dissociation, and agonists will tend to stabilize particular conformations. Because desensitized nAChR conformations have a higher affinity for agonist, nAChRs will increasingly adopt desensitized conformations in response to chronic nicotine exposure (Quick and Lester 2002; Picciotto et al. 2008). It has been suggested that receptor desensitization may be a major contributor to the upregulation observed in chronic users of nicotine-containing products (Benowitz 2008). As the usage patterns of regular smokers result in persisting levels of circulating nicotine over the course of the day, nAChR desensitization occurs in response to ongoing occupancy of CNS nAChRs. For example, working with 11 tobacco-dependent individuals using a radiotracer developed to image α4β2* nAChRs with positron-emission tomography (PET), it was found that regular smokers approach complete saturation of CNS α4β2* nAChRs throughout the day (Brody et al. 2006). After 2 days of abstaining from smoking, participants' cravings were reduced only once nAChRs were again nearly saturated. It is important to note that rates of desensitization are not necessarily equivalent across all nAChR subunit combinations. For example, inclusion of the α5 subunit into α4β2* nAChRs decreases the extent of desensitization (Bailey et al. 2010).

When considering the implications of receptor upregulation and desensitization for symptoms of nicotine withdrawal, kinetics characterizing the two dynamics may present critical processes by which withdrawal signaling occurs. In fact, nAChRs recover from desensitization on the scale of seconds to hours (Gentry and Lukas 2002). Additionally, studies have demonstrated that upregulation of nAChR expression can persist for several days following termination of chronic nicotine (Pietila et al. 1998; Staley et al. 2006). Accordingly, as the rates of desensitization and recovery from receptor upregulation during nicotine withdrawal are incongruent, a physiological landscape is established in which there are more receptors expressed than would be present in nicotine-naïve individuals, and these receptors are less responsive to agonist binding. When nicotine levels are low or absent, the nAChRs recover from desensitization, leading to potentially overactive cholinergic signaling. This process represents a major factor in the neurobiology of nicotine withdrawal (Dani and Heinemann 1996). Considering this dynamic, Gould and colleagues (Gould et al. 2012) have proposed a pharmacological strategy of treating hypersensitive nAChR systems with compounds that maintain receptor desensitization, or diminish cholinergic signaling while avoiding upregulation and activation, to alleviate the negative impact of withdrawal.

5 Anatomical Structures and Circuits Implicated in Nicotine Withdrawal

Given the diversity of symptoms manifested during nicotine withdrawal, a constellation of anatomical structures is likely involved in its etiology. Different anatomical structures in the CNS express distinct populations of nAChRs. The variety of nAChRs, combined with the specificity determined by afferent/efferent projections, produces the distinct neurochemistry and signaling that underlies the withdrawal syndrome. Evaluating changes in behavior relative to WT controls upon removal of particular nAChR subunits has been crucial to characterizing the contributions that each subunit makes to the withdrawal syndrome. If a withdrawal symptom exhibited by wild-type mice is absent in mice that lack functional expression of a specific nAChR subunit in a particular CNS structure, the specific subunit is likely critical to the withdrawal symptom in question. By performing these kinds of experiments, neuronal circuits contributing to the withdrawal syndrome will continue to be defined, and viable targets to enable the development of more specific therapeutics will be identified. Thus far, several brain structures have been already implicated in the withdrawal syndrome.

5.1 Dopaminergic System

The mesocorticolimbic dopamine (DA) system has been identified as critical to the development of addictive behaviors, and DA signaling is involved in reward-based reinforcement of drug-derived behaviors. While the DAergic VTA has been widely implicated in the rewarding aspects of addictive drugs, it has also been shown to participate in the signaling of aversion and lack of expected reward (Schultz et al. 1998;

Ungless et al. 2004; Tobler et al. 2007; De Biasi and Dani 2011). Dopaminergic deficits in the mesolimbic pathway, particularly in the nucleus accumbens, are among the neurochemical mechanisms underlying the symptoms of nicotine withdrawal (Hildebrand et al. 1998; Carboni et al. 2000; Rada et al. 2001). Such deficiencies in accumbal dopaminergic transmission are believed to contribute to the aversive anhedonic or dysphoric state experienced during nicotine abstinence (Koob and Le Moal 2008; Zhang et al. 2012). A reduction in basal DA levels was observed to linger for at least 5 days in mice following 12 weeks of chronic nicotine treatment (Zhang et al. 2012). This change in DA activity during withdrawal was correlated with a reduced modulatory influence by β2* nAChRs over DA release in the NAcc. What causes the hypodopaminergic state associated with withdrawal is not yet clear. However, increased inhibitory input to the VTA is a potential mechanism, as GABA input is sufficient to suppress burst firing even with excitatory inputs intact (Lobb et al. 2010; Jalabert et al. 2011). GABAergic inputs arrive at the VTA from the substantia nigra pars reticulata, nucleus accumbens, ventral pallidum/globus pallidus, laterodorsal tegmentum, pedunculopontine nuclei, diagonal band of Broca, bed nucleus of the stria terminalis, and the caudal tip of the VTA, termed the rostromedial tegmental nucleus (RMTg) (Geisler and Zahm 2005; Jhou et al. 2009; Kaufling et al. 2009). Activity in the RMTg increases upon exposure to aversive stimuli and decreases upon exposure to reward stimuli, representing inversely correlated signaling patterns (Hong et al. 2011).

5.2 Habenular Complex and Interpeduncular Nucleus

The habenular complex, comprising the lateral (LHb) and medial (MHb) nuclei, is a diencephalic, epithalamic structure located ventrally along the dorsal third ventricle (Dani and De Biasi 2013). The habenula has been demonstrated to participate in the signaling underlying fear, anxiety, depression, and stress (Viswanath et al. 2013). The LHb sends excitatory glutamatergic projections to the RMTg and is a major component of the circuit underlying negative reward (Dani and De Biasi 2013). Additionally, the cholinoceptive cellular population within the medial habenula (MHb) has been associated with the aversive effects of nicotine (Fowler et al. 2011). The most prominent efferent output from the MHb is to the IPN via the fasciculus retroflexus, forming the MHb–IPN axis. Through this projection, the MHb releases acetylcholine (ACh), substance P (SP), and glutamate onto the IPN (Zhao-Shea et al. 2013). The MHb also hosts the expression of norepinephrine (NE), serotonin, ATP, and several neuropeptides (Dao et al. 2014). The IPN is located in the ventral midbrain, with the VTA and median raphe nucleus located dorsally.

Mice chronically treated with nicotine exhibit symptoms of withdrawal upon microinjection of mecamylamine into the MHb and IPN, but fail to manifest withdrawal symptoms when mecamylamine is injected to the cortex, hippocampus, or VTA (Salas et al. 2009). Furthermore, lidocaine infusion into the MHb to inhibit signaling significantly blocks the manifestations of somatic symptoms resulting from both mecamylamine-precipitated and spontaneous induction of nicotine withdrawal (Zhao-Shea et al. 2013). Following precipitated nicotine withdrawal in WT mice, there are elevated markers of neuronal activation in GABAergic IPN cells (Zhao-Shea et al. 2013). The involvement of the IPN in the manifestation of nicotine withdrawal symptoms was demonstrated by expression of channelrhodopsin (ChR2), a light-driven excitatory ion channel, in the GABAergic cells of the IPN (Zhao-Shea et al. 2013). Additionally, infusion of an α3β4* nAChR-selective antagonist into the IPN elicited somatic withdrawal signs in nicotine-naïve mice and an even greater number of total symptoms in mice chronically treated with nicotine (Zhao-Shea et al. 2013). Considered together, these data strongly implicate the MHb–IPN circuit in the nicotine withdrawal syndrome. As previously discussed, the α5, α3, and β4 nAChR subunits participate in withdrawal symptomatology. They are each densely expressed in the MHb and/or IPN (Salas et al. 2003b, 2004) and are potential targets for treatments of the nicotine withdrawal syndrome (Grady et al. 2009).

5.3 Hippocampus

The hippocampus is a crucial structure for the manifestation of cognitive deficits during nicotine withdrawal. Contextual FC is dependent upon hippocampal signaling (Davis et al. 2005), and during spontaneous withdrawal from nicotine, mice exhibit deficient contextual FC relative to saline-treated controls (Davis et al. 2005). Administration of nicotine to mice undergoing withdrawal ameliorates this deficit (Davis et al. 2005). This effect was not attributable simply to a lowered threshold of fear responses inherent to the effects of nicotine because cued FC was equivalent between nicotine-treated and nicotine-naïve groups. Similar deficits of contextual fear learning were observed when a nAChR antagonist (dihydro-β-erythroidine) was infused into the hippocampus of WT mice chronically treated with nicotine (Davis and Gould 2009). The work also implicated β2* nAChRs in the manifestation of memory-associated deficits during nicotine withdrawal. Furthermore, chronic nicotine treatment upregulated a variety of nAChRs in the hippocampus, and the duration of β2 subunit upregulation relative to other nAChR subunits most closely corresponded to the total duration of memory-associated withdrawal symptoms (Gould et al. 2012).

5.4 Extended Amygdala

The bed nucleus of the stria terminalis (BNST), central nucleus of the amygdala (CeA), and shell of the nucleus accumbens (NAc-Sh) form the extended amygdala, an anatomically and neurochemically interconnected system located in the basal forebrain (Smith and Aston-Jones 2008). The system has been implicated in stress-related components of drug withdrawal and is a site of interaction between corticotropin-releasing factor (CRF) and NE transmission. Nicotine withdrawal leads to an increase in CRF in the CeA, and blockade of CRF1 receptors diminishes nicotine withdrawal-induced anxiety-like behavior (George et al. 2007). The CRF/CRF1 receptor signaling in the CeA also mediates nicotine withdrawal-induced increases in nociceptive sensitivity in rats that are dependent on nicotine (Baiamonte et al. 2014).

Given the polygenic nature of nicotine addiction, many other mechanisms and many other brain areas are likely to influence the manifestations of withdrawal. Preclinical research is increasingly benefitting from the information coming from genetic studies and pharmacogenetic trials, and the speed of discovery is destined to increase in the coming years.

6 Summary and Research Moving Forward

Individuals undergoing nicotine withdrawal experience both affective and somatic symptoms beginning between 4 and 24 h after ceasing intake. The syndrome is most severe in the first week, but it can persist for longer periods of time. During this time, relapse is incentivized by the ability of nicotine to alleviate or abolish withdrawal symptoms. In addition, there are cognitive deficits that manifest during nicotine withdrawal, including difficulty concentrating, increased reaction times in tasks requiring sustained attention, and impaired episodic and working memory (Myers et al. 2008; Wesnes et al. 2013). Knowledge of the molecular mechanisms that govern the emergence and intensity of withdrawal symptoms, and elucidation of the genetic variants associated with successful smoking cessation, will facilitate the development of personalized treatments.

New techniques in neuroscience are enabling researchers to ask previously unapproachable questions regarding which genes, circuits, and neurochemical systems mediate and modulate different aspects of addiction and withdrawal. Using transgenic mouse lines or viral delivery, designer receptors [DREADDs (Rogan and Roth 2011)] and light-driven ion channels [opsins (Yizhar et al. 2011)] can be expressed in particular anatomical structures or in particular neuronal types. These genetically targeted receptors and ion channels can be used to control the activities of specific circuits and populations of neurons in freely behaving mice. When combined with receptor KO mice, or with the delivery of nAChRs derived from specific human SNPs, these techniques enable researchers to manipulate neuronal activity while monitoring the dynamics of withdrawal-related behaviors.

Genetic studies of nicotine dependence and smoking cessation have identified several risk factors using GWAS, candidate gene approaches, and pharmacogenetic analyses. These studies provide targets that can be validated with large population samples and across ethnicities. Once putative genes are identified, hypotheses of the functional roles of such candidate genes can be tested in preclinical animal models. The functional consequences of some SNPs can be addressed relatively easily if the

SNPs are non-synonymous coding variants. The best example is provided by rs16969968, the SNP that leads to an aspartic acid for an asparagine substitution (D398N) in *CHRNA5* (Jackson et al. 2010). The impact of the mutation that defines the risk allele has been validated repeatedly in vitro and in animal models (Kuryatov et al. 2011; Morel et al. 2014). For many of the SNPs, however, there is no change in protein sequence, and therefore, it is harder to formulate clear functional hypothesis. Candidate SNPs can have a multitude of biochemical functions, such as altering DNA methylation, histone modification, or accessibility of DNA to transcription factor binding that can impact when, where, and how much a gene, and its protein, is expressed. Fortunately, analytical tools, such as the Encyclopedia of DNA Elements (ENCODE), can help the design of experiments that explore disease-related variants located within non-coding regions (Siggens and Ekwall 2014). Increasing attention is being paid to the functional analysis of rare variants, as common variants can explain only a small percentage of the variance in smoking-related phenotypes. More work is necessary at the bench and in the clinic, as a collection of genes likely operates collectively to predispose or protect individuals to or from nicotine addiction and withdrawal. Ultimately, this increased capability to comprehensively characterize the etiology of withdrawal symptoms will inform more competent and efficient drug design. The future undoubtedly holds greater development of pharmacological nicotine cessation therapeutics.

Acknowledgments *Funding*: This work was supported in part by the following NIH grants: DA024385, U19CA148127 (MDB); DA036572 (JAD & MDB) and DA009411 (JAD).

References

American Psychiatric Association (2013) Diagnostic and statistical manual of mental disorders: DSM-5. American Psychiatric Association, Washington

Baiamonte BA, Valenza M, Roltsch EA, Whitaker AM, Baynes BB, Sabino V, Gilpin NW (2014) Nicotine dependence produces hyperalgesia: role of corticotropin-releasing factor-1 receptors (CRF1Rs) in the central amygdala (CeA). Neuropharmacology 77:217–223

Bailey KR, Rustay NR, Crawley JN (2006) Behavioral phenotyping of transgenic and knockout mice: practical concerns and potential pitfalls. ILAR J 47:124–131

Bailey CD, De Biasi M, Fletcher PJ, Lambe EK (2010) The nicotinic acetylcholine receptor alpha5 subunit plays a key role in attention circuitry and accuracy. J Neurosci 30:9241–9252

Baker TB, Weiss RB, Bolt D, von Niederhausern A, Fiore MC, Dunn DM, Piper ME, Matsunami N, Smith SS, Coon H, McMahon WM, Scholand MB, Singh N, Hoidal JR, Kim SY, Leppert MF, Cannon DS (2009) Human neuronal acetylcholine receptor A5-A3-B4 haplotypes are associated with multiple nicotine dependence phenotypes. Nicotine Tob Res 11:785–796

Benowitz NL (2008) Neurobiology of nicotine addiction: implications for smoking cessation treatment. Am J Med 121:S3–S10

Benwell ME, Balfour DJ, Anderson JM (1988) Evidence that tobacco smoking increases the density of (-)-[3H]nicotine binding sites in human brain. J Neurochem 50:1243–1247

Bloom AJ, Martinez M, Chen LS, Bierut LJ, Murphy SE, Goate A (2013) CYP2B6 non-coding variation associated with smoking cessation is also associated with differences in allelic expression, splicing, and nicotine metabolism independent of common amino-acid changes. PLoS One 8:e79700

Booker T, Butt CM, Wehner JM, Heinemann SF, Collins AC (2007) Decreased anxiety-like behavior in beta3 nicotinic receptor subunit knockout mice. Pharmacol Biochem Behav 87:146–157

Bordia T, Hrachova M, Chin M, McIntosh JM, Quik M (2012) Varenicline is a potent partial agonist at alpha6beta2* nicotinic acetylcholine receptors in rat and monkey striatum. J Pharmacol Exp Ther 342:327–334

Brody AL, Mandelkern MA, London ED, Olmstead RE, Farahi J, Scheibal D, Jou J, Allen V, Tiongson E, Chefer SI, Koren AO, Mukhin AG (2006) Cigarette smoking saturates brain alpha 4 beta 2 nicotinic acetylcholine receptors. Arch Gen Psychiatry 63:907–915

Carboni E, Bortone L, Giua C, Di Chiara G (2000) Dissociation of physical abstinence signs from changes in extracellular dopamine in the nucleus accumbens and in the prefrontal cortex of nicotine dependent rats. Drug Alcohol Depend 58:93–102

Chen LS, Bloom AJ, Baker TB, Smith SS, Piper ME, Martinez M, Saccone N, Hatsukami D, Goate A, Bierut L (2014) Pharmacotherapy effects on smoking cessation vary with nicotine metabolism gene (CYP2A6). Addiction 109:128–137

Conti DV, Lee W, Li D, Liu J, Van Den Berg D, Thomas PD, Bergen AW, Swan GE, Tyndale RF, Benowitz NL, Lerman C (2008) Nicotinic acetylcholine receptor beta2 subunit gene implicated in a systems-based candidate gene study of smoking cessation. Hum Mol Genet 17:2834–2848

Crawley JN (1996) Unusual behavioral phenotypes of inbred mouse strains. Trends Neurosci 19:181–182 (discussion 188–189)

Crawley JN (2008) Behavioral phenotyping strategies for mutant mice. Neuron 57:809–818

Crawley J, Belknap JK, Collins A, Crabbe JC, Frankel W, Henderson N, Hitzemann RJ, Maxson SC, Miner LL, Silva AJ (1997) Behavioral phenotypes of inbred mouse strains: implications and recommendations for molecular studies. Psychopharmacology 132:107–124

Cryan JF, Markou A, Lucki I (2002) Assessing antidepressant activity in rodents: recent developments and future needs. Trends Pharmacol Sci 23:238–245

Cui C, Booker TK, Allen RS, Grady SR, Whiteaker P, Marks MJ, Salminen O, Tritto T, Butt CM, Allen WR, Stitzel JA, McIntosh JM, Boulter J, Collins AC, Heinemann SF (2003) The beta3 nicotinic receptor subunit: a component of alpha-conotoxin MII-binding nicotinic acetylcholine receptors that modulate dopamine release and related behaviors. J Neurosci 23:11045–11053

Dani JA, De Biasi M (2013) Mesolimbic dopamine and habenulo-interpeduncular pathways in nicotine withdrawal. Cold Spring Harb perspect Med 3:012138

Dani JA, Heinemann S (1996) Molecular and cellular aspects of nicotine abuse. Neuron 16:905–908

Dao DQ, Perez EE, Teng Y, Dani JA, De Biasi M (2014) Nicotine enhances excitability of medial habenular neurons via facilitation of neurokinin signaling. J Neurosci 34:4273–4284

Dash B, Lukas RJ, Li MD (2014) A signal peptide missense mutation associated with nicotine dependence alters alpha2*-nicotinic acetylcholine receptor function. Neuropharmacology 79:715–725

Davis JA, Gould TJ (2008) Associative learning, the hippocampus, and nicotine addiction. Curr Drug Abuse Rev 1:9–19

Davis JA, Gould TJ (2009) Hippocampal nAChRs mediate nicotine withdrawal-related learning deficits. Eur Neuropsychopharmacol 19:551–561

Davis JA, James JR, Siegel SJ, Gould TJ (2005) Withdrawal from chronic nicotine administration impairs contextual fear conditioning in C57BL/6 mice. J Neurosci 25:8708–8713

De Biasi M, Dani JA (2011) Reward, addiction, withdrawal to nicotine. Annu Rev Neurosci 34:105

De Biasi M, Salas R (2008) Influence of neuronal nicotinic receptors over nicotine addiction and withdrawal. Exp Biol Med 233:917–929

Dokal I, Pagliuca A, Deenmamode M, Mufti GJ, Lewis SM (1989) Development of polycythaemia vera in a patient with myelofibrosis. Eur J Haematol 42:96–98

Exley R, Clements MA, Hartung H, McIntosh JM, Cragg SJ (2008) Alpha6-containing nicotinic acetylcholine receptors dominate the nicotine control of dopamine neurotransmission in nucleus accumbens. Neuropsychopharmacology 33:2158–2166

Fanselow MS, Poulos AM (2005) The neuroscience of mammalian associative learning. Annu Rev Psychol 56:207–234

Fenster CP, Whitworth TL, Sheffield EB, Quick MW, Lester RA (1999) Upregulation of surface alpha4beta2 nicotinic receptors is initiated by receptor desensitization after chronic exposure to nicotine. J Neurosci 19:4804–4814

Fowler CD, Lu Q, Johnson PM, Marks MJ, Kenny PJ (2011) Habenular [agr] 5 nicotinic receptor subunit signalling controls nicotine intake. Nature 471:597–601

Fowler CD, Tuesta L, Kenny PJ (2013) Role of alpha5* nicotinic acetylcholine receptors in the effects of acute and chronic nicotine treatment on brain reward function in mice. Psychopharmacology 229:503–513

Frahm S, Ślimak MA, Ferrarese L, Santos-Torres J, Antolin-Fontes B, Auer S, Filkin S, Pons S, Fontaine J-F, Tsetlin V (2011) Aversion to nicotine is regulated by the balanced activity of β4 and α5 nicotinic receptor subunits in the medial habenula. Neuron 70:522–535

Furberg H, Kim Y, Dackor J, Boerwinkle E, Franceschini N, Ardissino D, Bernardinelli L, Mannucci P, Mauri F, Merlini P, Absher D, Assimes T, Fortmann S, Iribarren C, Knowles J, Quertermous T, Ferrucci L, Tanaka T, Bis J, Furberg C, Haritunians T, McKnight B, Psaty B, Taylor K, Thacker E, Almgren P, Groop L, Ladenvall C, Boehnke M, Jackson A, Mohlke K, Stringham H, Tuomilehto J, Benjamin E, Hwang S, Levy D, Preis S, Vasan R, Duan J, Gejman P, Levinson D, Sanders A, Shi J, Lips E, McKay J, Agudo A, Barzan L, Bencko V, Benhamou S, Castellsague X, Canova C, Conway D, Fabianova E, Foretova L, Janout V, Healy C, Holcátová I, Kjaerheim K, Lagiou P, Lissowska J, Lowry R, Macfarlane T, Mates D, Richiardi L, Rudnai P, Szeszenia-Dabrowska N, Zaridze D, Znaor A, Lathrop M, Brennan P, Bandinelli S, Frayling T, Guralnik J, Milaneschi Y, Perry J, Altshuler D, Elosua R, Kathiresan S, Lucas G, Melander O, O'Donnell C, Salomaa V, Schwartz S, Voight B, Penninx B, Smit J, Vogelzangs N, Boomsma D, de Geus E, Vink J, Willemsen G, Chanock S, Gu F, Hankinson S, Hunter D, Hofman A, Tiemeier H, Uitterlinden A, van Duijn C, Walter S, Chasman D, Everett B, Paré G, Ridker P, Li M, Maes H, Audrain-McGovern J, Posthuma D, Thornton L, Lerman C, Kaprio J, Rose J, Ioannidis J, Kraft P, Lin D, Sullivan P (2010) Genome-wide meta-analyses identify multiple loci associated with smoking behavior. Nat Genet 42:441–447

Gangitano D, Salas R, Teng Y, Perez E, De Biasi M (2009) Progesterone modulation of alpha5 nAChR subunits influences anxiety-related behavior during estrus cycle. Genes Brain Behav 8:398–406

Geisler S, Zahm DS (2005) Afferents of the ventral tegmental area in the rat-anatomical substratum for integrative functions. J Comp Neurol 490:270–294

Gentry CL, Lukas RJ (2002) Regulation of nicotinic acetylcholine receptor numbers and function by chronic nicotine exposure. Curr Drug Targets CNS Neurol Disord 1:359–385

George O, Ghozland S, Azar MR, Cottone P, Zorrilla EP, Parsons LH, O'Dell LE, Richardson HN, Koob GF (2007) CRF-CRF1 system activation mediates withdrawal-induced increases in nicotine self-administration in nicotine-dependent rats. Proc Natl Acad Sci USA 104:17198–17203

Gold AB, Lerman C (2012) Pharmacogenetics of smoking cessation: role of nicotine target and metabolism genes. Hum Genet 131:857–876

Gould TJ, Portugal GS, Andre JM, Tadman MP, Marks MJ, Kenney JW, Yildirim E, Adoff M (2012) The duration of nicotine withdrawal-associated deficits in contextual fear conditioning parallels changes in hippocampal high affinity nicotinic acetylcholine receptor upregulation. Neuropharmacology 62:2118–2125

Govind AP, Vezina P, Green WN (2009) Nicotine-induced upregulation of nicotinic receptors: underlying mechanisms and relevance to nicotine addiction. Biochem Pharmacol 78:756–765

Grabus SD, Martin BR, Batman AM, Tyndale RF, Sellers E, Damaj MI (2005) Nicotine physical dependence and tolerance in the mouse following chronic oral administration. Psychopharmacology 178:183–192

Grady SR, Salminen O, Laverty DC, Whiteaker P, McIntosh JM, Collins AC, Marks MJ (2007) The subtypes of nicotinic acetylcholine receptors on dopaminergic terminals of mouse striatum. Biochem Pharmacol 74:1235–1246

Grady SR, Moretti M, Zoli M, Marks MJ, Zanardi A, Pucci L, Clementi F, Gotti C (2009) Rodent habenulo-interpeduncular pathway expresses a large variety of uncommon nAChR subtypes, but only the α3β4 and α3β3β4 subtypes mediate acetylcholine release. J Neurosci 29:2272–2282

Grady SR, Drenan RM, Breining SR, Yohannes D, Wageman CR, Fedorov NB, McKinney S, Whiteaker P, Bencherif M, Lester HA, Marks MJ (2010) Structural differences determine the relative selectivity of nicotinic compounds for native alpha 4 beta 2*-, alpha 6 beta 2*-, alpha 3 beta 4*- and alpha 7-nicotine acetylcholine receptors. Neuropharmacology 58:1054–1066

Graham AW, Schultz TK, Mayo-Smith MF, Ries RK, Wilford B (2007) Principles of addiction medicine. Lippincott Williams & Wilkins, Philadelphia

Gu DF, Hinks LJ, Morton NE, Day IN (2000) The use of long PCR to confirm three common alleles at the CYP2A6 locus and the relationship between genotype and smoking habit. Ann Hum Genet 64:383–390

Hall W, Madden P, Lynskey M (2002) The genetics of tobacco use: methods, findings and policy implications. Tob Control 11:119–124

Haller G, Druley T, Vallania FL, Mitra RD, Li P, Akk G, Steinbach JH, Breslau N, Johnson E, Hatsukami D, Stitzel J, Bierut LJ, Goate AM (2012) Rare missense variants in CHRNB4 are associated with reduced risk of nicotine dependence. Hum Mol Genet 21:647–655

Hartz SM, Short SE, Saccone NL, Culverhouse R, Chen L, Schwantes-An TH, Coon H, Han Y, Stephens SH, Sun J, Chen X, Ducci F, Dueker N, Franceschini N, Frank J, Geller F, Gubjartsson D, Hansel NN, Jiang C, Keskitalo-Vuokko K, Liu Z, Lyytikainen LP, Michel M, Rawal R, Rosenberger A, Scheet P, Shaffer JR, Teumer A, Thompson JR, Vink JM, Vogelzangs N, Wenzlaff AS, Wheeler W, Xiao X, Yang BZ, Aggen SH, Balmforth AJ, Baumeister SE, Beaty T, Bennett S, Bergen AW, Boyd HA, Broms U, Campbell H, Chatterjee N, Chen J, Cheng YC, Cichon S, Couper D, Cucca F, Dick DM, Foroud T, Furberg H, Giegling I, Gu F, Hall AS, Hallfors J, Han S, Hartmann AM, Hayward C, Heikkila K, Hewitt JK, Hottenga JJ, Jensen MK, Jousilahti P, Kaakinen M, Kittner SJ, Konte B, Korhonen T, Landi MT, Laatikainen T, Leppert M, Levy SM, Mathias RA, McNeil DW, Medland SE, Montgomery GW, Muley T, Murray T, Nauck M, North K, Pergadia M, Polasek O, Ramos EM, Ripatti S, Risch A, Ruczinski I, Rudan I, Salomaa V, Schlessinger D, Styrkarsdottir U, Terracciano A, Uda M, Willemsen G, Wu X, Abecasis G, Barnes K, Bickeboller H, Boerwinkle E, Boomsma DI, Caporaso N, Duan J, Edenberg HJ, Francks C, Gejman PV, Gelernter J, Grabe HJ, Hops H, Jarvelin MR, Viikari J, Kahonen M, Kendler KS, Lehtimaki T, Levinson DF, Marazita ML, Marchini J, Melbye M, Mitchell BD, Murray JC, Nothen MM, Penninx BW, Raitakari O, Rietschel M, Rujescu D, Samani NJ, Sanders AR, Schwartz AG, Shete S, Shi J, Spitz M, Stefansson K, Swan GE, Thorgeirsson T, Volzke H, Wei Q, Wichmann HE, Amos CI, Breslau N, Cannon DS, Ehringer M, Grucza R, Hatsukami D, Heath A, Johnson EO, Kaprio J, Madden P, Martin NG, Stevens VL, Stitzel JA, Weiss RB, Kraft P, Bierut LJ (2012) Increased genetic vulnerability to smoking at CHRNA5 in early-onset smokers. Arch Gen Psychiatry 69:854–860

Hecht SS, Hochalter JB, Villalta PW, Murphy SE (2000) 2′-Hydroxylation of nicotine by cytochrome P450 2A6 and human liver microsomes: formation of a lung carcinogen precursor. Proc Natl Acad Sci USA 97:12493–12497

Heishman SJ, Kleykamp BA, Singleton EG (2010) Meta-analysis of the acute effects of nicotine and smoking on human performance. Psychopharmacology 210:453–469

Henderson BJ, Srinivasan R, Nichols WA, Dilworth CN, Gutierrez DF, Mackey ED, McKinney S, Drenan RM, Richards CI, Lester HA (2014) Nicotine exploits a COPI-mediated process for chaperone-mediated up-regulation of its receptors. J Gen Physiol 143:51–66

Hildebrand BE, Nomikos GG, Hertel P, Schilstrom B, Svensson TH (1998) Reduced dopamine output in the nucleus accumbens but not in the medial prefrontal cortex in rats displaying a mecamylamine-precipitated nicotine withdrawal syndrome. Brain Res 779:214–225

Hong S, Jhou TC, Smith M, Saleem KS, Hikosaka O (2011) Negative reward signals from the lateral habenula to dopamine neurons are mediated by rostromedial tegmental nucleus in primates. J Neurosci 31:11457–11471

Jackson KJ, Martin BR, Changeux J-P, Damaj MI (2008) Differential role of nicotinic acetylcholine receptor subunits in physical and affective nicotine withdrawal signs. J Pharmacol Exp Ther 325:302–312

Jackson KJ, McIntosh JM, Brunzell DH, Sanjakdar SS, Damaj MI (2009) The role of alpha6-containing nicotinic acetylcholine receptors in nicotine reward and withdrawal. J Pharmacol Exp Ther 331:547–554

Jackson KJ, Marks MJ, Vann RE, Chen X, Gamage TF, Warner JA, Damaj MI (2010) Role of α5 nicotinic acetylcholine receptors in pharmacological and behavioral effects of nicotine in mice. J Pharmacol Exp Ther 334:137–146

Jackson KJ, Sanjakdar SS, Muldoon PP, McIntosh JM, Damaj MI (2013) The alpha3beta4* nicotinic acetylcholine receptor subtype mediates nicotine reward and physical nicotine withdrawal signs independently of the alpha5 subunit in the mouse. Neuropharmacology 70:228–235

Jalabert M, Bourdy R, Courtin J, Veinante P, Manzoni OJ, Barrot M, Georges F (2011) Neuronal circuits underlying acute morphine action on dopamine neurons. Proc Natl Acad Sci USA 108:16446–16450

Jhou TC, Fields HL, Baxter MG, Saper CB, Holland PC (2009) The rostromedial tegmental nucleus (RMTg), a GABAergic afferent to midbrain dopamine neurons, encodes aversive stimuli and inhibits motor responses. Neuron 61:786–800

Kaufling J, Veinante P, Pawlowski SA, Freund-Mercier MJ, Barrot M (2009) Afferents to the GABAergic tail of the ventral tegmental area in the rat. J Comp Neurol 513:597–621

Kenny PJ, Markou A (2001) Neurobiology of the nicotine withdrawal syndrome. Pharmacol Biochem Behav 70:531–549

King DP, Paciga S, Pickering E, Benowitz NL, Bierut LJ, Conti DV, Kaprio J, Lerman C, Park PW (2012) Smoking cessation pharmacogenetics: analysis of varenicline and bupropion in placebo-controlled clinical trials. Neuropsychopharmacology 37:641–650

Koob GF, Le Moal M (2008) Addiction and the brain antireward system. Annu Rev Psychol 59:29–53

Krackow S, Vannoni E, Codita A, Mohammed AH, Cirulli F, Branchi I, Alleva E, Reichelt A, Willuweit A, Voikar V, Colacicco G, Wolfer DP, Buschmann JU, Safi K, Lipp HP (2010) Consistent behavioral phenotype differences between inbred mouse strains in the intellicage. Genes Brain Behav 9:722–731

Kubota T, Nakajima-Taniguchi C, Fukuda T, Funamoto M, Maeda M, Tange E, Ueki R, Kawashima K, Hara H, Fujio Y, Azuma J (2006) CYP2A6 polymorphisms are associated with nicotine dependence and influence withdrawal symptoms in smoking cessation. Pharmacogenomics J 6:115–119

Kuryatov A, Berrettini W, Lindstrom J (2011) Acetylcholine receptor (AChR) alpha5 subunit variant associated with risk for nicotine dependence and lung cancer reduces (alpha4beta2)(2) alpha5 AChR function. Mol Pharmacol 79:119–125

Lalonde R, Strazielle C (2008) Relations between open-field, elevated plus-maze, and emergence tests as displayed by C57/BL6J and BALB/c mice. J Neurosci Methods 171:48–52

Lee AM, Jepson C, Hoffmann E, Epstein L, Hawk LW, Lerman C, Tyndale RF (2007) CYP2B6 genotype alters abstinence rates in a bupropion smoking cessation trial. Biol Psychiatry 62:635–641

Leeb J, Tamse A (1985) The use of calcium hydroxide in endodontic therapy. Refuat Hashinayim 3:3–12

Lessov CN, Martin NG, Statham DJ, Todorov AA, Slutske WS, Bucholz KK, Heath AC, Madden PA (2004) Defining nicotine dependence for genetic research: evidence from Australian twins. Psychol Med 34:865–879

Li MD (2008) Identifying susceptibility loci for nicotine dependence: 2008 update based on recent genome-wide linkage analyses. Hum Genet 123:119–131

Lobb CJ, Wilson CJ, Paladini CA (2010) A dynamic role for GABA receptors on the firing pattern of midbrain dopaminergic neurons. J Neurophysiol 104:403–413

Lotfipour S, Byun JS, Leach P, Fowler CD, Murphy NP, Kenny PJ, Gould TJ, Boulter J (2013) Targeted Deletion of the Mouse α2 Nicotinic Acetylcholine Receptor Subunit Gene (Chrna2) Potentiates Nicotine-Modulated Behaviors. J Neurosci 33:7728–7741

Luo S, Kulak JM, Cartier GE, Jacobsen RB, Yoshikami D, Olivera BM, McIntosh JM (1998) alpha-conotoxin AuIB selectively blocks alpha3 beta4 nicotinic acetylcholine receptors and nicotine-evoked norepinephrine release. J Neurosci 18:8571–8579

Malin DH, Lake JR, Carter VA, Cunningham JS, Hebert KM, Conrad DL, Wilson OB (1994) The nicotinic antagonist mecamylamine precipitates nicotine abstinence syndrome in the rat. Psychopharmacology 115:180–184

Mao D, Perry DC, Yasuda RP, Wolfe BB, Kellar KJ (2008) The alpha4beta2alpha5 nicotinic cholinergic receptor in rat brain is resistant to up-regulation by nicotine in vivo. J Neurochem 104:446–456

Marks MJ, Burch JB, Collins AC (1983) Effects of chronic nicotine infusion on tolerance development and nicotinic receptors. J Pharmacol Exp Ther 226:817–825

Matsuo N, Takao K, Nakanishi K, Yamasaki N, Tanda K, Miyakawa T (2010) Behavioral profiles of three C57BL/6 substrains. Front Behav Neurosci 4:29

Morel C, Fattore L, Pons S, Hay YA, Marti F, Lambolez B, De Biasi M, Lathrop M, Fratta W, Maskos U, Faure P (2014) Nicotine consumption is regulated by a human polymorphism in dopamine neurons. Mol Psychiatry 19:930–936

Myers CS, Taylor RC, Moolchan ET, Heishman SJ (2008) Dose-related enhancement of mood and cognition in smokers administered nicotine nasal spray. Neuropsychopharmacology 33:588–598

Nugent KL, Million-Mrkva A, Backman J, Stephens SH, Reed RM, Kochunov P, Pollin TI, Shuldiner AR, Mitchell BD, Hong LE (2014) Familial aggregation of tobacco use behaviors among amish men. Nicotine Tob Res 16:923–930

O'Dell LE, Khroyan TV (2009) Rodent models of nicotine reward: what do they tell us about tobacco abuse in humans? Pharmacol Biochem Behav 91:481–488

Paolini M, De Biasi M (2011) Mechanistic insights into nicotine withdrawal. Biochem Pharmacol 82:996–1007

Pergadia ML, Heath AC, Martin NG, Madden PA (2006) Genetic analyses of DSM-IV nicotine withdrawal in adult twins. Psychol Med 36:963–972

Pergadia ML, Agrawal A, Loukola A, Montgomery GW, Broms U, Saccone SF, Wang JC, Todorov AA, Heikkila K, Statham DJ, Henders AK, Campbell MJ, Rice JP, Todd RD, Heath AC, Goate AM, Peltonen L, Kaprio J, Martin NG, Madden PA (2009) Genetic linkage findings for DSM-IV nicotine withdrawal in two populations. Am J Med Genet B Neuropsychiatr Genet 150b:950–959

Perkins KA, Lerman C, Mercincavage M, Fonte CA, Briski JL (2009) Nicotinic acetylcholine receptor beta2 subunit (CHRNB2) gene and short-term ability to quit smoking in response to nicotine patch. Cancer Epidemiol Biomarkers Prev 18:2608–2612

Perry DC, Mao D, Gold AB, McIntosh JM, Pezzullo JC, Kellar KJ (2007) Chronic nicotine differentially regulates alpha6- and beta3-containing nicotinic cholinergic receptors in rat brain. J Pharmacol Exp Ther 322:306–315

Picciotto MR, Addy NA, Mineur YS, Brunzell DH (2008) It is not "either/or": activation and desensitization of nicotinic acetylcholine receptors both contribute to behaviors related to nicotine addiction and mood. Prog Neurobiol 84:329–342

Pietila K, Lahde T, Attila M, Ahtee L, Nordberg A (1998) Regulation of nicotinic receptors in the brain of mice withdrawn from chronic oral nicotine treatment. Naunyn Schmiedebergs Arch Pharmacol 357:176–182

Portugal GS, Wilkinson DS, Turner JR, Blendy JA, Gould TJ (2012) Developmental effects of acute, chronic, and withdrawal from chronic nicotine on fear conditioning. Neurobiol Learn Mem 97:482–494

Quick MW, Lester RA (2002) Desensitization of neuronal nicotinic receptors. J Neurobiol 53:457–478

Rada P, Jensen K, Hoebel BG (2001) Effects of nicotine and mecamylamine-induced withdrawal on extracellular dopamine and acetylcholine in the rat nucleus accumbens. Psychopharmacology 157:105–110

Rennard SI, Daughton DM (2014) Smoking cessation. Clin Chest Med 35:165–176

Rezvani K, Teng Y, Pan Y, Dani JA, Lindstrom J, Garcia Gras EA, McIntosh JM, De Biasi M (2009) UBXD4, a UBX-containing protein, regulates the cell surface number and stability of alpha3-containing nicotinic acetylcholine receptors. J Neurosci 29:6883–6896

Rezvani K, Teng Y, De Biasi M (2010) The ubiquitin–proteasome system regulates the stability of neuronal nicotinic acetylcholine receptors. J Mol Neurosci 40:177–184

Robbins TW (2002) The 5-choice serial reaction time task: behavioural pharmacology and functional neurochemistry. Psychopharmacology 163:362–380

Robinson JD, Lam CY, Minnix JA, Wetter DW, Tomlinson GE, Minna JD, Chen TT, Cinciripini PM (2007) The DRD2 TaqI-B polymorphism and its relationship to smoking abstinence and withdrawal symptoms. Pharmacogenomics J 7:266–274

Rogan SC, Roth BL (2011) Remote control of neuronal signaling. Pharmacol Rev 63:291–315

Saccone NL, Saccone SF, Hinrichs AL, Stitzel JA, Duan W, Pergadia ML, Agrawal A, Breslau N, Grucza RA, Hatsukami D, Johnson EO, Madden PA, Swan GE, Wang JC, Goate AM, Rice JP, Bierut LJ (2009) Multiple distinct risk loci for nicotine dependence identified by dense coverage of the complete family of nicotinic receptor subunit (CHRN) genes. Am J Med Genet B Neuropsychiatr Genet 150b:453–466

Salas R, Orr-Urtreger A, Broide RS, Beaudet A, Paylor R, De Biasi M (2003a) The nicotinic acetylcholine receptor subunit alpha 5 mediates short-term effects of nicotine in vivo. Mol Pharmacol 63:1059–1066

Salas R, Pieri F, Fung B, Dani JA, De Biasi M (2003b) Altered anxiety-related responses in mutant mice lacking the beta4 subunit of the nicotinic receptor. J Neurosci 23:6255–6263

Salas R, Pieri F, De Biasi M (2004) Decreased signs of nicotine withdrawal in mice null for the β4 nicotinic acetylcholine receptor subunit. J Neurosci 24:10035–10039

Salas R, Main A, Gangitano D, De Biasi M (2007) Decreased withdrawal symptoms but normal tolerance to nicotine in mice null for the α7 nicotinic acetylcholine receptor subunit. Neuropharmacology 53:863–869

Salas R, Sturm R, Boulter J, De Biasi M (2009) Nicotinic receptors in the habenulo-interpeduncular system are necessary for nicotine withdrawal in mice. J Neurosci 29:3014–3018

Sallette J, Bohler S, Benoit P, Soudant M, Pons S, Le Novere N, Changeux JP, Corringer PJ (2004) An extracellular protein microdomain controls up-regulation of neuronal nicotinic acetylcholine receptors by nicotine. J Biol Chem 279:18767–18775

Sarginson JE, Killen JD, Lazzeroni LC, Fortmann SP, Ryan HS, Schatzberg AF, Murphy GM Jr (2011) Markers in the 15q24 nicotinic receptor subunit gene cluster (CHRNA5-A3-B4) predict severity of nicotine addiction and response to smoking cessation therapy. Am J Med Genet B Neuropsychiatr Genet 156b:275–284

Schultz W, Tremblay L, Hollerman JR (1998) Reward prediction in primate basal ganglia and frontal cortex. Neuropharmacology 37:421–429

Shoaib M, Bizarro L (2005) Deficits in a sustained attention task following nicotine withdrawal in rats. Psychopharmacology 178:211–222

Siggens L, Ekwall K (2014) Epigenetics, chromatin and genome organization: recent advances from the ENCODE project. J Intern Med 276:201–214

Sigurdsson T, Doyere V, Cain CK, LeDoux JE (2007) Long-term potentiation in the amygdala: a cellular mechanism of fear learning and memory. Neuropharmacology 52:215–227

Smith RJ, Aston-Jones G (2008) Noradrenergic transmission in the extended amygdala: role in increased drug-seeking and relapse during protracted drug abstinence. Brain Struct Funct 213:43–61

Srinivasan R, Pantoja R, Moss FJ, Mackey ED, Son CD, Miwa J, Lester HA (2011) Nicotine up-regulates alpha4beta2 nicotinic receptors and ER exit sites via stoichiometry-dependent chaperoning. J Gen Physiol 137:59–79

Staley JK, Krishnan-Sarin S, Cosgrove KP, Krantzler E, Frohlich E, Perry E, Dubin JA, Estok K, Brenner E, Baldwin RM (2006) Human tobacco smokers in early abstinence have higher levels of β2* nicotinic acetylcholine receptors than nonsmokers. J Neurosci 26:8707–8714

Stoker AK, Olivier B, Markou A (2012) Role of α7-and β4-containing nicotinic acetylcholine receptors in the affective and somatic aspects of nicotine withdrawal: studies in knockout mice. Behav Genet 42:423–436

Thanos P, Delis F, Rosko L, Volkow ND (2013) Passive response to stress in adolescent female and adult male mice after intermittent nicotine exposure in adolescence. J Addict Res Ther Suppl 6:007

Thorgeirsson TE, Gudbjartsson DF, Surakka I, Vink JM, Amin N, Geller F, Sulem P, Rafnar T, Esko T, Walter S, Gieger C, Rawal R, Mangino M, Prokopenko I, Magi R, Keskitalo K, Gudjonsdottir IH, Gretarsdottir S, Stefansson H, Thompson JR, Aulchenko YS, Nelis M, Aben KK, den Heijer M, Dirksen A, Ashraf H, Soranzo N, Valdes AM, Steves C, Uitterlinden AG, Hofman A, Tonjes A, Kovacs P, Hottenga JJ, Willemsen G, Vogelzangs N, Doring A, Dahmen N, Nitz B, Pergadia ML, Saez B, De Diego V, Lezcano V, Garcia-Prats MD, Ripatti S, Perola M, Kettunen J, Hartikainen AL, Pouta A, Laitinen J, Isohanni M, Huei-Yi S, Allen M, Krestyaninova M, Hall AS, Jones GT, van Rij AM, Mueller T, Dieplinger B, Haltmayer M, Jonsson S, Matthiasson SE, Oskarsson H, Tyrfingsson T, Kiemeney LA, Mayordomo JI, Lindholt JS, Pedersen JH, Franklin WA, Wolf H, Montgomery GW, Heath AC, Martin NG, Madden PA, Giegling I, Rujescu D, Jarvelin MR, Salomaa V, Stumvoll M, Spector TD, Wichmann HE, Metspalu A, Samani NJ, Penninx BW, Oostra BA, Boomsma DI, Tiemeier H, van Duijn CM, Kaprio J, Gulcher JR, McCarthy MI, Peltonen L, Thorsteinsdottir U, Stefansson K (2010) Sequence variants at CHRNB3-CHRNA6 and CYP2A6 affect smoking behavior. Nat Genet 42:448–453

Tobler PN, O'Doherty JP, Dolan RJ, Schultz W (2007) Reward value coding distinct from risk attitude-related uncertainty coding in human reward systems. J Neurophysiol 97:1621

Tumkosit P, Kuryatov A, Luo J, Lindstrom J (2006) Beta3 subunits promote expression and nicotine-induced up-regulation of human nicotinic alpha6* nicotinic acetylcholine receptors expressed in transfected cell lines. Mol Pharmacol 70:1358–1368

Turner JR, Castellano LM, Blendy JA (2011) Parallel anxiolytic-like effects and upregulation of neuronal nicotinic acetylcholine receptors following chronic nicotine and varenicline. Nicotine Tob Res 13:41–46

Uhl GR, Liu QR, Drgon T, Johnson C, Walther D, Rose JE (2007) Molecular genetics of nicotine dependence and abstinence: whole genome association using 520,000 SNPs. BMC Genet 8:10

Ungless MA, Magill PJ, Bolam JP (2004) Uniform inhibition of dopamine neurons in the ventral tegmental area by aversive stimuli. Science 303:2040–2042

Viswanath H, Carter AQ, Baldwin PR, Molfese DL, Salas R (2013) The medial habenula: still neglected. Font Hum Neurosci 7:931

Wahlsten D (2010) Mouse behavioral testing: how to use mice in behavioral neuroscience. Academic Press, London

Wang F, Nelson ME, Kuryatov A, Olale F, Cooper J, Keyser K, Lindstrom J (1998) Chronic nicotine treatment up-regulates human alpha3 beta2 but not alpha3 beta4 acetylcholine receptors stably transfected in human embryonic kidney cells. J Biol Chem 273:28721–28732

Wang S, van der Vaart AD, Xu Q, Seneviratne C, Pomerleau OF, Pomerleau CS, Payne TJ, Ma JZ, Li MD (2014) Significant associations of CHRNA2 and CHRNA6 with nicotine dependence in European American and African American populations. Hum Genet 133:575–586

Wesnes KA, Edgar CJ, Kezic I, Salih HM, de Boer P (2013) Effects of nicotine withdrawal on cognition in a clinical trial setting. Psychopharmacology 229:133–140

Xian H, Scherrer JF, Madden PA, Lyons MJ, Tsuang M, True WR, Eisen SA (2003) The heritability of failed smoking cessation and nicotine withdrawal in twins who smoked and attempted to quit. Nicotine Tob Res 5:245–254

Xian H, Scherrer JF, Madden PA, Lyons MJ, Tsuang M, True WR, Eisen SA (2005) Latent class typology of nicotine withdrawal: genetic contributions and association with failed smoking cessation and psychiatric disorders. Psychol Med 35:409–419

Xu W, Gelber S, Orr-Urtreger A, Armstrong D, Lewis RA, Ou CN, Patrick J, Role L, De Biasi M, Beaudet AL (1999) Megacystis, mydriasis, and ion channel defect in mice lacking the alpha3 neuronal nicotinic acetylcholine receptor. Proc Natl Acad Sci USA 96:5746–5751

Yizhar O, Fenno LE, Davidson TJ, Mogri M, Deisseroth K (2011) Optogenetics in neural systems. Neuron 71:9–34

Zhang L, Dong Y, Doyon WM, Dani JA (2012) Withdrawal from chronic nicotine exposure alters dopamine signaling dynamics in the nucleus accumbens. Biol Psychiatry 71:184–191

Zhao-Shea R, Liu L, Soll LG, Improgo MR, Meyers EE, McIntosh JM, Grady SR, Marks MJ, Gardner PD, Tapper AR (2011) Nicotine-mediated activation of dopaminergic neurons in distinct regions of the ventral tegmental area. Neuropsychopharmacology 36:1021–1032

Zhao-Shea R, Liu L, Pang X, Gardner PD, Tapper AR (2013) Activation of GABAergic neurons in the interpeduncular nucleus triggers physical nicotine withdrawal symptoms. Curr Biol 23:2327–2335

Neurobiological Bases of Cue- and Nicotine-induced Reinstatement of Nicotine Seeking: Implications for the Development of Smoking Cessation Medications

Astrid K. Stoker and Athina Markou

Abstract A better understanding of the neurobiological factors that contribute to relapse to smoking is needed for the development of efficacious smoking cessation medications. Reinstatement procedures allow the preclinical assessment of several factors that contribute to relapse in humans, including re-exposure to nicotine via tobacco smoking and the presentation of stimuli that were previously associated with nicotine administration (i.e., conditioned stimuli). This review provides an integrated discussion of the results of animal studies that used reinstatement procedures to assess the efficacy of pharmacologically targeting various neurotransmitter systems in attenuating the cue- and nicotine-induced reinstatement of nicotine seeking. The results of these animal studies have increased our understanding of the neurobiological processes that mediate the conditioned effects of stimuli that trigger reinstatement to nicotine seeking. Thus, these findings provide important insights into the neurobiological substrates that modulate relapse to tobacco smoking in humans and the ongoing search for novel efficacious smoking cessation medications.

Keywords Animal models · Nicotine · Cue-induced reinstatement · Nicotine-induced reinstatement · Drug seeking · Medication development

A.K. Stoker · A. Markou
Department of Psychiatry, School of Medicine, University of California San Diego, La Jolla, CA, USA

A.K. Stoker
Department of Pharmacology and Systems Therapeutics, Icahn School of Medicine at Mount Sinai, New York, NY, USA

A. Markou (✉)
Department of Psychiatry, M/C 0603, School of Medicine, University of California San Diego, 9500 Gilman Drive, La Jolla, CA 92093-0603, USA
e-mail: amarkou@ucsd.edu

D.J.K. Balfour and M.R. Munafò (eds.), *The Neuropharmacology of Nicotine Dependence*, Current Topics in Behavioral Neurosciences 24,
DOI 10.1007/978-3-319-13482-6_5

Contents

1 Introduction

The negative impact of tobacco consumption on health remains one of the most urgent health issues (Alberg et al. 2014) because the global number of tobacco smokers continues to steadily increase (Ng et al. 2014). Although the health risks associated with tobacco smoking are greatly reduced by the cessation of tobacco consumption, fewer than 5 % of all quit attempts result in lifelong abstinence from tobacco smoking (Hughes et al. 2004). The vast majority of current smokers have considered or undertaken at least one quit attempt. Specifically, a recent survey of tobacco smokers in the USA reported that two-thirds of the respondents wished to quit their harmful habit and that over half of the respondents had undertaken one quit attempt in the previous year (Centers for Disease Control and Prevention 2010). However, despite the currently available smoking cessation medications and behavioral intervention treatments, an estimated half of smokers failed to quit their habit, despite multiple quit attempts during their lifetime (Centers for Disease Control and Prevention 2000). Relapse to tobacco seeking is thus one of the most defining features of tobacco dependence. To decrease relapse rates, the identification of novel pharmacological targets is needed. The development of these novel pharmacological targets in the treatment of tobacco dependence requires the use of procedures in experimental animals that mimic aspects of relapse in humans.

Nicotine is the main psychoactive ingredient in tobacco (Stolerman and Jarvis 1995). Therefore, experimental animal research that focuses on identifying pharmacological targets for novel smoking aids primarily assesses the effects of nicotine. The reinstatement procedure is one of the most widely used tools for screening the effects of pharmacological compounds on "relapse" to nicotine seeking in animals. The reinstatement of nicotine-seeking behavior is broadly defined as the continuation of the behavioral response that previously resulted in nicotine delivery after noncontingent exposure to nicotine (nicotine-induced reinstatement) or the presentation of stimuli (e.g., cue light illumination and sounds associated with activation of the pump that delivers nicotine) that were previously associated with nicotine administration (cue-induced reinstatement) after a period of abstinence. In the reinstatement procedure, animals are allowed to self-administer nicotine for a prolonged period of time before undergoing extinction training. During extinction training, the animals are placed in the chambers where they were previously allowed to intravenously self-administer nicotine, while nicotine and its conditioned cues are withheld. Alternatively to extinction training, animals may also undergo a period of forced abstinence. During this period of abstinence, also referred to as the incubation of nicotine seeking (Abdolahi et al. 2010; Bedi et al. 2011), the animals remain in their home cages during the withdrawal period. Nicotine seeking in animals can then be reinstated by manipulations that have been associated with relapse in human smokers. Conditions that induce relapse to tobacco smoking in humans include tobacco smoking itself, conditioned cues, and stress (Doherty et al. 1995). In parallel to humans, rodent studies have shown that noncontingent nicotine administration (Dravolina et al. 2007), conditioned cues (Paterson et al. 2005), and stress (Bruijnzeel et al. 2009) all induce the reinstatement of nicotine-seeking behavior (see chapters entitled Behavioral Mechanisms Underlying Nicotine Reinforcement and The Role of Mesoaccumbens Dopamine in Nicotine Dependence; this volume). Importantly, the overlap in factors that induce relapse in humans and reinstatement in experimental animals suggests good etiological validity for the reinstatement model.

Predictive validity for the reinstatement procedure in terms of pharmacological isomorphism (Geyer and Markou 1995) is provided by various studies that identified compounds that attenuate both relapse in humans and the reinstatement of nicotine seeking in animals. For example, the opioid receptor antagonist naltrexone facilitated smoking cessation and reduced relapse rates in humans (Epstein and King 2004; King 2002; King et al. 2012, 2013; King and Meyer 2000) and effectively attenuated the reinstatement of nicotine seeking in rats (Liu et al. 2009). Another example is the cannabinoid CB_1 receptor antagonist rimonabant, which similarly decreased relapse to tobacco consumption in humans and reinstatement to nicotine seeking in rats (Cahill and Ussher 2011; Diergaarde et al. 2008; Forget et al. 2009). In contrast, however, two smoking cessation medications that are currently approved by the United States Food and Drug Administration (FDA) produced mixed results in studies of reinstatement of nicotine seeking in rats. Specifically, bupropion enhanced cue-induced reinstatement of nicotine seeking (Liu et al. 2008), while varenicline attenuated nicotine-induced, but not cue-induced, reinstatement of nicotine seeking (O'Connor et al. 2010). This variability in findings may reflect the different neurocircuits that probably underlie different aspects of the reinstatement of nicotine seeking. Results from reinstatement studies of drug seeking for other psychostimulant drugs, including cocaine, support the postulation that distinct neurocircuits are involved in the regulation of diverse aspects of the reinstatement of drug seeking (for reviews, see Kalivas and

McFarland 2003; See et al. 2003). The sections below briefly describe several neurocircuits that regulate diverse aspects of nicotine dependence, followed by discussions of the various neurotransmitter systems that are of interest in studies of the reinstatement of nicotine seeking.

2 Neurocircuits of Interest in Studies of the Reinstatement of Nicotine Seeking

2.1 Mesolimbic System

Dopaminergic projections from the ventral tegmental area (VTA) to nucleus accumbens (NAc) are critically involved in mediating the reinforcing and motivational properties of nicotine (Gerasimov et al. 2000; Laviolette and van der Kooy 2004). Nicotine directly activates these dopaminergic projections through β2-containing nicotinic acetylcholine receptors (nAChRs; see chapters entitled Structure of Neuronal Nicotinic Receptors and Genetics of Smoking Behaviour; volume 23) on VTA dopaminergic neurons (Mameli-Engvall et al. 2006) and indirectly through α7 nAChRs located on VTA glutamatergic neurons (Mansvelder and McGehee 2000). Pharmacological compounds that provide low levels of stimulation of dopaminergic projections to the NAc may therefore result in decreased nicotine-seeking behavior. In fact, the efficacy of the nicotinic receptor partial agonist varenicline, one of the few FDA-approved smoking cessation aids, in attenuating relapse to smoking in humans is assumed to partially result from low stimulatory actions at β2-containing nAChRs located on dopaminergic projections from the VTA to NAc (West et al. 2008). Alternatively, increased glutamatergic neurotransmission through the activation of excitatory α7 nAChRs, located on glutamatergic terminals that synapse on VTA dopamine neurons, activates dopaminergic projections from the VTA to NAc (Mansvelder and McGehee 2000). Moreover, a recent study by Gipson et al. (2013) demonstrated that glutamatergic neurotransmission in the NAc contributed to the cue-induced reinstatement of nicotine-seeking behavior, supporting an important role for the mesolimbic neurocircuit in the reinforcing and motivational effects of nicotine and the reinstatement of nicotine seeking. Findings from rat studies indicated changes in glutamatergic neurotransmission in the NAc during early nicotine withdrawal (i.e., 24 h after the cessation of nicotine self-administration; Knackstedt et al. 2009; Liechti et al. 2007). Subsequent studies by Gipson et al. (2013) revealed that these changes in NAc glutamatergic synaptic plasticity persist when the withdrawal period is extended to 2 weeks. Moreover, the cue-induced reinstatement procedure induced simultaneous increases in glutamatergic neurotransmission in the NAc and nicotine seeking (Gipson et al. 2013), further indicating the importance of glutamatergic neurotransmission in the mediation of cue-induced reinstatement.

2.2 Habenulo-interpeduncular Circuit

Opposite to the role of the corticolimbic neurocircuit in the reinforcing effects of nicotine, the habenulo-interpeduncular circuit appears to be important in mediating the aversive effects of nicotine (Fowler et al. 2011, 2013; Frahm et al. 2011). The medial habenula became of interest in the study of nicotine dependence when genetic linkage studies found that genetic variation in the gene cluster that encodes the α3, α5, and β4 nAChR subunits, which are densely located in the habenular circuit (De Biasi and Salas 2008), was associated with lung cancer and nicotine dependence in humans (Saccone et al. 2007, 2010). Mice null for the α5 nAChR subunit vigorously self-administered very high concentrations of nicotine (Fowler et al. 2011). Moreover, re-expression of the α5 nAChR subunit in the medial habenula in these knockout mice returned their increased self-administration rates of high nicotine doses back to rates observed in wild-type mice (Fowler et al. 2011), highlighting the involvement of the habenula-interpeduncular pathway in the high nicotine intake in these knockout mice. The effects of null mutation of the α5 nAChR subunit on nicotine consumption were explained in a later study by Fowler and colleagues. This study suggested that the reward-suppressing effects that high nicotine doses induce in wild-type mice were absent in α5 knockout mice (Fowler et al. 2013). Similarly, increased activity of β4 nAChR subunits in mice, in which the *CHRNA5-CHRNA3-CHRNB4* gene cluster was co-expressed with a bacterial artificial chromosome, resulted in the consumption of markedly less nicotine in a no-choice bottle procedure, presumably because of the aversion to nicotine that these mice exhibit in the conditioned place aversion procedure (Frahm et al. 2011; see also chapter entitled Genetics of Smoking Behaviour; volume 23). Lentiviral expression of the D398N α5 variant, which has been genetically associated with nicotine dependence and lung cancer in humans (Falvella et al. 2009; Hung et al. 2008; Wang et al. 2009), in the medial habenula reversed the aversion to nicotine in these mice (Frahm et al. 2011). These studies suggest the involvement of habenular α5 and β4 nAChR subunits in the aversive effects of nicotine, but the role of the habenulo-interpeduncular circuit in the reinstatement of nicotine-seeking behavior remains to be explored. Nevertheless, the somatic signs of nicotine withdrawal were attenuated in mice null for the both α5 and β4 nAChR subunits (Jackson et al. 2008; Salas et al. 2007; Stoker et al. 2012), suggesting that these nAChR subunits may be involved in mediating at least some of the aversive aspects of the nicotine withdrawal syndrome and that pharmacologically targeting these nAChR subunits may alleviate the negative withdrawal symptoms that ultimately result in the reinstatement of nicotine seeking.

2.3 *Insular Cortex and Dorsal Striatum*

Compared with the mesolimbic and habenulo-interpeduncular neurocircuits, the insula and dorsal striatum are thought to be recruited in later stages of drug dependence when drug-taking behavior becomes more habitual (for reviews, see Everitt et al. 2008; Naqvi and Bechara 2009). These brain structures are therefore particularly interesting in the study of drug reinstatement. The insular cortex became of interest in the study of relapse to tobacco smoking when Naqvi and colleagues reported that damage to the insular cortex facilitated spontaneous smoking cessation in humans, presumably by relieving symptoms of craving (Naqvi et al. 2007). Similarly, pharmacological or electrical inactivation of the insula attenuated the cue- and nicotine-induced reinstatement of nicotine-seeking behavior in animals (Forget et al. 2010a; Pushparaj et al. 2013). Subsequent studies provided additional evidence that the insula critically regulates different aspects of dependence on various psychostimulants and is particularly important in mediating the effects of conditioned cues associated with drug craving (Abdolahi et al. 2010; Contreras et al. 2007, 2012; Forget et al. 2010a; Hollander et al. 2008; Scott and Hiroi 2011). The insula projects to the dorsal striatum, a brain region implicated in the habitual and compulsive aspects of drug dependence. Interestingly, a recent case report described a patient in whom a lesion of the dorsal striatum resulted in the attenuation of nicotine intake (Muskens et al. 2012), similar to the effects of lesions of the insula on smoking cessation. In animals, studies of the involvement of the dorsal striatum in drug seeking have primarily focused on psychostimulant drugs other than nicotine, most notably cocaine. Lesions or pharmacological inactivation of the dorsal striatum in rats attenuated cocaine-seeking behavior (Fuchs et al. 2006; Fucile et al. 1997; Fung and Richard 1994; Gabriele and See 2011). The effects of conditioned cues associated with cocaine were shown to be mediated by dopaminergic neurotransmission in the dorsal striatum in animals (Ito et al. 2002) and humans (Volkow et al. 2006).

3 Role of Various Neurotransmitter Systems in Cue- and Nicotine-induced Reinstatement of Nicotine-seeking Behavior

3.1 *Acetylcholine*

All smoking cessation medications that are currently approved by the FDA (i.e., nicotine replacement therapy, varenicline, and bupropion) have some affinity at nAChRs (Coe et al. 2005; Slemmer et al. 2000; Thompson and Hunter 1998). While these medications are ineffective in approximately 80 % of smokers who attempt to quit smoking (Gonzales et al. 2006; Hughes et al. 2003; Jorenby et al. 2006), nAChRs have remained key targets in the research and development of novel pharmacotherapies for smoking cessation. Bupropion, a medication widely used for smoking cessation, was initially used as an antidepressant therapy and primarily acts as a dopamine and norepinephrine reuptake inhibitor. Interestingly, bupropion was also found to act as an nAChR antagonist after it was marketed as a smoking cessation medication (Slemmer et al. 2000). bupropion has had mixed effects on smoking cessation rates in humans (Hurt et al. 1997) and actually enhanced the cue-induced reinstatement of nicotine seeking in rats (Liu et al. 2008). These results suggest that bupropion may have limited efficacy at treating the full spectrum of relapse and may primarily alleviate depressive-like symptoms during withdrawal (Cryan et al. 2003; Paterson et al. 2007).

Varenicline, a partial agonist at α4β2-containing nAChRs (Coe et al. 2005), was synthesized after studies suggested the crucial involvement of β2 nAChRs in nicotine dependence (Maskos et al. 2005; Picciotto 1998). Preclinical studies that assessed the effects of varenicline on the reinstatement of nicotine-seeking behavior demonstrated mixed effects on cue- and nicotine-induced reinstatement. Specifically, the nicotine-induced reinstatement of nicotine seeking was attenuated by varenicline in rats (see Fig. 1b adapted from O'Connor et al. 2010), similar to the effects of varenicline in humans. In contrast, varenicline did not affect the cue-induced reinstatement of nicotine-seeking behavior in rats (O'Connor et al. 2010; Wouda et al. 2011, see Fig. 1a adapted from Wouda et al. 2011), and higher doses of varenicline even enhanced cue-induced reinstatement (Wouda ct al. 2011). Interestingly, varenicline attenuated the cue-induced reinstatement of nicotine seeking with a prolonged pretreatment time (Le Foll et al. 2012). When reinstatement was induced by both nicotine priming and the presentation of cues, varenicline also attenuated the reinstatement of nicotine seeking (O'Connor et al. 2010), presumably because of the attenuating effect of varenicline on nicotine-induced reinstatement. Altogether, preclinical studies on the effects of varenicline on nicotine seeking suggest the differential regulation of cue- and nicotine-induced reinstatement. Furthermore, Liu (2014) demonstrated that α7 nAChR antagonism with methyllycaconitine (MLA) but not α4β2-containing nAChR antagonism with dihydro-β-erythroidine (DHβE) reduced the cue-induced reinstatement of nicotine seeking. The results with varenicline and DHβE suggest that α4β2-containing nAChRs may be involved in the regulation of nicotine-induced, but not cue-induced reinstatement of nicotine seeking. In contrast, α7 nAChRs may be involved in mediating the cue-induced reinstatement of nicotine seeking. The enhancement of cue-induced reinstatement that results from administration of higher doses of varenicline (Wouda et al. 2011) may be explained by the activation of α7 nAChRs by varenicline, which acts as a full agonist at these receptors (Mihalak et al. 2006). This interpretation is further supported by results from the study by Liu (2014), which showed that α7 nAChR blockade attenuated cue-induced reinstatement. Further support for the involvement of α7 nAChRs in nicotine seeking reinstated by the presentation of conditioned cues, but not nicotine, is provided by the results of a study that reported that TAT-α7-pep2, a protein that interferes with the function of the α7nAChR–NMDA receptor complex, reduced the cue-induced reinstatement of

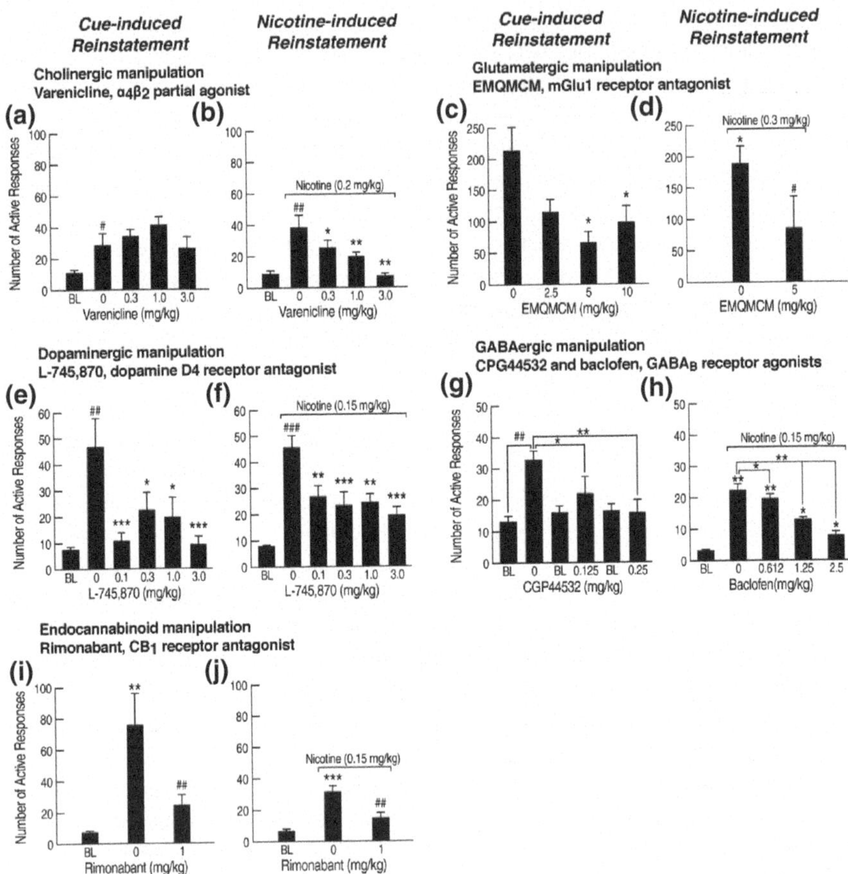

Fig. 1 Effects of manipulating various neurotransmitter systems on cue- and nicotine-induced reinstatement of nicotine-seeking behavior. The pharmacological targeting of a wide range of neurotransmitter systems attenuated reinstatement to nicotine seeking induced by the presentation of conditioned cues (*left panel*) and nicotine priming (*right panel*). Positive allosteric modulation of α4β2 nAChRs with varenicline had no effect on the cue-induced reinstatement of nicotine seeking (**a**, modified with permission from O'Connor et al. 2010), while varenicline attenuated the nicotine-induced reinstatement of nicotine seeking (**b**, modified with permission from O'Connor et al. 2010). The mGlu1 receptor antagonist EMQMCM attenuated both the cue-induced (**c**, modified with permission from Dravolina et al. 2007) and nicotine-induced (**d**, modified with permission from Dravolina et al. 2007) reinstatement of nicotine-seeking behavior. The dopamine D_4 receptor antagonist L-745,870 also attenuated the cue-induced (**e**, modified with permission from Yan et al. 2013) and nicotine-induced (**f**, modified with permission from Yan et al. 2013) reinstatement of nicotine seeking. The $GABA_B$ receptor agonists CPG44532 (**g**, modified with permission from Paterson et al. 2005) and baclofen (**h**, modified with permission from Fattore et al. 2009) similarly reduced the cue- and nicotine-induced reinstatement of nicotine seeking, respectively. Finally, the CB_1 receptor antagonist rimonabant attenuated the cue-induced (**i,** modified with permission from Forget et al. 2009) and nicotine-induced (**j**, modified with permission from Forget et al. 2009) reinstatement of nicotine seeking

nicotine seeking while not affecting reinstatement induced by nicotine priming (Li et al. 2012).

In addition to the direct activation of nAChRs, cholinergic neurotransmission can be increased by the inhibition of acetylcholinesterase, the enzyme that metabolizes the endogenous nAChR ligand acetylcholine. Galantamine, which acts both as an acetylcholinesterase inhibitor and positive allosteric modulator at α7- and α4β2-containing nAChRs, reduced the cue-induced reinstatement of nicotine seeking, suggesting that acetylcholinesterase inhibitors may be effective tools in the prevention of nicotine reinstatement (Hopkins et al. 2012). The potential therapeutic value of acetylcholinesterase inhibitors was supported by a study that demonstrated that donezipil, which acts exclusively as an acetylcholinesterase inhibitor, attenuated nicotine-induced reinstatement (Kimmey et al. 2012). Combined, the findings with galantamine and donezipil suggest that acetylcholinesterase inhibitors can attenuate both cue- and nicotine-induced reinstatement.

In summary, the pharmacological targeting of acetylcholinergic neurotransmission has been one of the most lucrative avenues in the search for smoking cessation aids to date. However, cue- and nicotine-induced reinstatement appears to be differentially regulated by pharmacological compounds that act on α7- and α4β2-containing nAChRs, potentially limiting their efficacy as pharmacological targets for smoking cessation medication. Exploring the efficacy of pharmacologically targeting diverse nAChR subtypes, including α3-, α5-, and β4-containing nAChR subunits, may thus be an interesting avenue in the identification of novel, highly efficacious smoking cessation medications.

3.2 Glutamate

Glutamatergic neurotransmission is modulated by two different types of receptors: ionotropic glutamate (iGlu) receptors and metabotropic glutamate (mGlu) receptors. iGlu receptors are located postsynaptically and modulate fast glutamatergic neurotransmission. Nicotine self-administration resulted in changes in iGlu and mGlu receptor levels, which likely contributed to the cue-induced reinstatement of nicotine-seeking behavior (Gipson et al. 2013; Liechti et al. 2007). Moreover, pharmacologically targeting iGlu receptors with the NMDA receptor antagonists ifenprodil and acamprosate attenuated the cue-induced reinstatement of nicotine-seeking behavior (Gipson et al. 2013; Pechnick et al. 2011). The development of novel pharmacotherapies for the treatment of drug dependence, however, has focused primarily on mGlu receptors because of the side effects of iGlu receptor antagonists in humans (for review, see Gass and Olive 2008). Metabotropic glutamate receptors have attracted much interest in recent years as targets for novel therapeutics in the treatment of nicotine dependence (Markou 2007). Compared with iGlu receptors, the activity of mGlu receptors is more slow acting and modulatory, presumably resulting in a reduced side effect profile. During nicotine withdrawal, presynaptic mGlu2/3 receptors were downregulated in the VTA and

NAc, and mGlu2/3 receptor activation in these brain areas induced by the agonist LY379268 attenuated the cue-induced reinstatement of nicotine seeking (Liechti et al. 2007). N-acetylcysteine, a compound that has been suggested to increase the glutamatergic tone of presynaptic mGlu2/3 receptors (Kupchik et al. 2012), similarly attenuated nicotine reinstatement elicited by environmental cues (Ramirez-Nino et al. 2013). Furthermore, the blockade of postsynaptic mGlu5 receptors (Bespalov et al. 2005) and mGlu1 receptors (see Fig. 1c adapted from Dravolina et al. 2007) decreased cue-induced reinstatement. Specifically, nicotine seeking was attenuated by administration of the mGlu5 receptor antagonist 2-methyl-6-(phenylethynyl)-pyridine hydrochloride (MPEP; Bespalov et al. 2005) or mGlu1 receptor antagonist 3-ethyl-2-methyl-quinolin-6-yl)-(4-methoxy-cyclohexyl)-methanone methanesulfonate (EMQMCM; Dravolina et al. 2007). In parallel to cue-induced reinstatement, EMQMCM also decreased the nicotine-induced reinstatement of nicotine-seeking behavior in rats (see Fig. 1d adapted from Dravolina et al. 2007). These results of experimental studies in animals on the role of mGlu receptors in nicotine reinstatement demonstrated that pharmacologically targeting glutamatergic neurotransmission effectively attenuates both cue- and nicotine-induced reinstatement and may attenuate relapse to tobacco smoking in humans. In fact, these experimental animal studies resulted in a Phase I clinical trial by Novartis that assessed the efficacy and safety of the mGlu5 receptor antagonist AFQ056 as a treatment option for voluntary smoking cessation. This clinical trial has been completed, but the results of the study have not yet been published (Clinicaltrials.gov 2007).

3.3 Dopamine

As described in Sects. 2.1 and 2.3, dopaminergic neurotransmission, which is mediated by G-protein-coupled dopamine receptors, in the mesolimbic circuit and striatum is critically involved in drug dependence (see chapter entitled The Role of Mesoaccumbens Dopamine in Nicotine Dependence; this volume). The cue-induced reinstatement of nicotine-seeking behavior can be attenuated by pharmacological compounds that decrease dopaminergic tone, including antagonists of dopamine D_1 and D_2 receptors (Liu et al. 2010), D_3 receptors (Khaled et al. 2010), and D_4 receptors (see Fig. 1e adapted from Yan et al. 2013). Consistent with these findings, a reduction of dopaminergic tone with the α-type peroxisome proliferator-activated receptor (PPAR-α) agonist clofibrate decreased cue-induced reinstatement in squirrel monkeys (Panlilio et al. 2012). These studies suggest that the pharmacological inhibition of dopaminergic neurotransmission consistently attenuates the cue-induced reinstatement of nicotine-seeking behavior. Furthermore, nicotine-induced reinstatement is similarly attenuated by dopamine D_3 and D_4 receptor agonists (Andreoli et al. 2003; Yan et al. 2013, see Fig. 1f adapted from Yan et al. 2013) and PPAR-α agonists (Mascia et al. 2011; Panlilio et al. 2012). The inhibition of dopaminergic neurotransmission, therefore, appears to be an interesting possibility in the identification for novel smoking cessation medication targets.

3.4 γ-Aminobutyric Acid (GABA)

GABA is the main inhibitory transmitter in the central nervous system. Inhibitory GABAergic activity attenuates dopaminergic mesocorticolimbic neurotransmission through GABA interneurons located in the VTA, medium spiny GABA neurons in the NAc, and GABAergic projections to the VTA from the NAc, ventral pallidum, and pedunculopontine tegmental nucleus (Klitenick et al. 1992). Inhibitory GABA receptors, therefore, would have to be activated by full agonists or positive allosteric modulators to decrease excitatory neurotransmission in the VTA which, as discussed above, generally attenuates the reinstatement of nicotine seeking. GABAergic neurotransmission is regulated through ionotropic $GABA_A$ and $GABA_C$ receptors and metabotropic $GABA_B$ receptors (Bormann 1986). Of these various GABA receptor subtypes, G-protein-coupled $GABA_B$ receptors are primarily of interest in the treatment of nicotine dependence (Li et al. 2014; Vlachou and Markou 2010). $GABA_B$ receptor activation induced by the $GABA_B$ receptor agonist CPG44532 attenuated the cue-induced reinstatement of nicotine seeking in rats (see Fig. 1g adapted from Paterson et al. 2005). Similar to the $GABA_B$ agonist, the $GABA_B$ receptor positive allosteric modulator BHF177 also decreased cue-induced reinstatement in rats (Vlachou et al. 2011). The effects of $GABA_B$ agonists on the nicotine-induced reinstatement of nicotine seeking have been less extensively explored. One study reported that baclofen decreased reinstatement induced by nicotine priming (see Fig. 1h adapted from Fattore et al. 2009). Furthermore, the $GABA_B$ receptor agonist baclofen was suggested to potentially facilitate smoking cessation in humans (Cousins et al. 2001). These studies indicate that $GABA_B$ receptors may be a promising target in the treatment of smoking cessation. Moreover, it has been proposed that $GABA_B$ positive allosteric modulators may be particularly effective in the treatment of nicotine dependence because of their modulatory actions at $GABA_B$ receptors that may result in an improved side effect profile and decreased development of tolerance to these compounds compared with $GABA_B$ full receptor agonists (Guery et al. 2007; Vlachou et al. 2011).

3.5 Endocannabinoids

Of the two endocannabinoid receptors cloned to date, CB_1 receptors are of primary interest in the treatment of dependence on drugs of abuse (Howlett et al. 2004) because these receptors are found on glutamatergic and GABAergic inputs to dopaminergic neurons (Gardner 2005). In contrast, CB_2 receptors are primarily localized on immune cells in both the central and peripheral nervous systems (Howlett 2002). As expected, the reinstatement of nicotine-seeking behavior was unaffected by the CB_2 receptor antagonist AM630 or CB_2 receptor agonist AM1241 (Gamaleddin et al. 2012b). CB_1 receptors located on presynaptic glutamatergic neurons in the VTA are hypothesized to decrease the inhibitory control that

GABAergic neurons exert on dopaminergic neurons (Schlicker and Kathmann 2001). Consequently, CB_1 receptor activation would result in the increased firing activity of VTA dopamine neurons (French 1997; French et al. 1997) and increased dopamine release in the NAc (Gardner and Vorel 1998; Tanda et al. 1997), suggesting therapeutic potential for CB_1 receptor antagonism in attenuating the reinstatement of nicotine seeking. Indeed, antagonism of the CB_1 receptor consistently attenuated the cue-induced reinstatement of nicotine-seeking behavior, demonstrated by the administration of rimonabant (Diergaarde et al. 2008; Forget et al. 2009, see Fig. 1i adapted from Forget et al. 2009), SR141716 (Cohen et al. 2005; de Vries et al. 2005), and AM404 (Gamaleddin et al. 2013) in rats. The $CB_{1/2}$ receptor agonist WIN 55,212-2 facilitated the cue-induced reinstatement of nicotine-seeking behavior (Gamaleddin et al. 2012a), presumably by activating CB_1 receptors. Furthermore, antagonism at CB_1 receptors attenuated nicotine-induced reinstatement in rats, demonstrated by the administration of rimonabant (see Fig. 1j adapted from Forget et al. 2009), AM251 (Shoaib 2008), and AM404 (Gamaleddin et al. 2013). Additionally, reinstatement induced by the combination of both cue presentation and nicotine priming was attenuated by administration of the CB_1 receptor antagonist AM251 (Shoaib 2008).

After various clinical trials assessed the efficacy of rimonabant as a smoking cessation medication, it was approved for this purpose in various European countries in 2006. Inopportunely, treatment of smoking cessation with rimonabant was halted in 2007 after reports of severe side effects that included anxiety and depression (Moreira and Crippa 2009). The development of pharmacological compounds that target the endocannabinoid system in smoking cessation is therefore currently directed toward developing compounds that indirectly target endocannabinoid neurotransmission, including anandamide transport inhibitors and fatty acid amid hydrolase (FAAH). Anandamide is one of the endogenous ligands that act at cannabinoid receptors (Giang and Cravatt 1997) and eliminated by reuptake into cells by anandamide transporters and subsequent hydrolysis by FAAH (Beltramo et al. 1997; Cravatt et al. 1996). The enhancement of endocannabinoid signaling by inhibiting the reuptake or hydrolysis of anandamide attenuated both cue- and nicotine-induced reinstatement of nicotine seeking (Forget et al. 2009; Gamaleddin et al. 2011). Notably, inhibiting the reuptake or hydrolysis of anandamide opposes the effects of CB_1 receptor antagonists. That is, CB_1 receptor antagonists attenuate endocannabinoid signaling, while FAAH inhibitors and anandamide transport inhibitors enhance endocannabinoid signaling. The same direction of effect (i.e., decrease) on the reinstatement of nicotine seeking by these seemingly opposing mechanisms may be due to the action of CB_1 receptor antagonists on neurocircuits that express endocannabinoid ligands other than anandamide (Scherma et al. 2008). Interestingly, compounds that target FAAH may exert dual actions on the reinstatement of nicotine seeking because FAAH also breaks down fatty acid amides that can activate PPAR-α (Fegley et al. 2005). As discussed above, the activation of PPAR-α decreases the cue-induced reinstatement of nicotine seeking, presumably by decreasing dopaminergic neurotransmission, further emphasizing the potential of FAAH-inhibiting compounds in treating the reinstatement of nicotine seeking.

3.6 Other Neurotransmitter Systems

Whereas the aforementioned neurotransmitter systems have been the most extensively explored as targets for pharmacotherapy in attenuating the reinstatement of nicotine seeking, several other neurotransmitter systems have been suggested in the development of novel smoking cessation aids. Serotonergic receptors, for example, modulate dopaminergic neurotransmission and were suggested as potential targets in smoking cessation medications (for review, see Fletcher et al. 2008). The attenuation of serotonergic neurotransmission by the 5-HT_{2C} receptor antagonists Ro60-0175 and locaserin decreased both nicotine- and cue-induced reinstatement (Fletcher et al. 2012; Higgins et al. 2012). Similarly, the modulation of noradrenergic neurotransmission was shown to effectively reduce both cue- and nicotine-induced reinstatement with the noradrenergic α1 receptor antagonist prazosin (Forget et al. 2010b) and β-blocker propranolol, supporting the involvement of noradrenergic neurotransmission in cue-induced reinstatement (Chiamulera et al. 2010). Finally, the T-type calcium channel antagonist TTA-A2 also attenuated both cue- and nicotine-induced reinstatement (Uslaner et al. 2010), potentially by modulating glutamatergic or dopaminergic neurotransmission (Uslaner et al. 2012). Tricyclic antidepressants, which decrease both serotonergic and noradrenergic neurotransmission, have been suggested to facilitate smoking cessation in humans (Edwards et al. 1989; Hall et al. 1998; Prochazka et al. 1998). However, the severe side effects of tricyclic antidepressants, including cardiovascular effects and the severity of overdose symptoms (Biggs et al. 1977; Roose et al. 1991), make these compounds unfavorable as smoking cessation aids. The reversible monoamine oxidase-A (MAO-A) inhibitor moclobemide, which reduces both serotonergic and noradrenergic neurotransmission similarly to tricyclic antidepressants but with a more favorable side effect profile (Stabl et al. 1989), attenuated smoking cessation in a study group of heavy smokers (Berlin et al. 1995). These results suggest that MAO-A inhibitors may be preferred over tricyclic antidepressants as smoking cessation aids. However, bupropion remains the main antidepressant used as a smoking cessation medication (Tables 1 and 2).

4 Concluding Remarks

Relapse is one of the hallmarks of tobacco dependence, but the currently available smoking cessation medications only prevent the occurrence of relapse in a small percentage of people who attempt to quit tobacco consumption. The limited

Table 1 Animal studies of the pharmacological targeting of various neurotransmitter systems in cue-induced reinstatement of nicotine seeking

Compound	Mechanism	Species/strain	Reinstatement	Effect on reinstatement	References
Acetylcholine					
DHβE	α4β2 nAChR antagonist	Sprague-Dawley rats	Cue-induced	–	Liu (2014)
Mecamylamine	Wide-spectrum nAChR antagonist	Sprague-Dawley rats	Cue-induced	↓	Liu et al. (2007)
MLA	α7 nAChR antagonist	Sprague-Dawley rats	Cue-induced	↓	Liu (2014)
TAT-α7-pep2	α7nAChR–NMDAR complex interfering protein	Long-Evans rats	Cue-induced	↓	Li et al. (2012)
varenicline	α4β2 nAChR partial agonist	Hooded Lister	Cue-induced	–	(O'Connor et al. 2010)
		Wistar rats	Cue-induced	–/↑	Wouda et al. (2011)
varenicline (long pre-treatment time)	α4β2 nAChR partial agonist	Long-Evans rats	Cue-induced	↓	Le Foll et al. (2012)
Glutamate					
EMQMCM	mGlu1 receptor antagonist	Wistar rats	Cue-induced	↓	Dravolina et al. (2007)
MPEP	mGlu5 receptor antagonist	Wistar rats	Cue-induced	↓	Bespalov et al. (2005)
LY379268	mGlu2/3 receptor agonist	Wistar rats	Cue-induced	↓	Liechti et al. (2007)
Acamprosate	NMDA receptor antagonist, $GABA_A$ receptor agonist	Sprague-Dawley rats	Cue-induced	↓	Pechnick et al. (2011)

(continued)

Table 1 (continued)

Compound	Mechanism	Species/strain	Reinstatement	Effect on reinstatement	References
N-acetylcysteine	Cysteine pro-drug	Wistar rats	Cue-induced	↓	Ramirez-Nino et al. (2013)
GABA					
Acamprosate	$GABA_A$ receptor agonist, NMDA receptor antagonist	Sprague-Dawley rats	Cue-induced	↓	Pechnick et al. (2011)
BHF177	$GABA_B$ receptor positive allosteric modulator	Wistar rats	Cue-induced	↓	Vlachou et al. (2011)
CPG44532	$GABA_B$ receptor agonist	Wistar rats	Cue-induced	↓	Paterson et al. (2005)
Dopamine					
BP897	Dopamine D_3 receptor agonist	Long-Evans rats	Cue-induced	–	Khaled et al. (2010)
Clofibrate	α-type peroxisome proliferator-activated receptor agonist	Squirrel monkeys	Cue-induced	↓	Panlilio et al. (2012)
Eticlopride	Dopamine D_2 receptor antagonist	Sprague-Dawley rats	Cue-induced	↓	(Liu et al. 2010)
L-745,870	Dopamine D_4 receptor antagonist	Long-Evans rats	Cue-induced	↓	(Yan et al. 2013)
SB277011-A	Dopamine D_3 receptor antagonist	Long-Evans rats	Cue-induced	↓	(Khaled et al. 2010)
SCH23390	Dopamine D_1 receptor antagonist	Sprague-Dawley rats	Cue-induced	↓	Liu et al. (2010)
Endocannabinoid					
AM404	CB_1 receptor antagonist	Long-Evans rats	Cue-induced	↓	Gamaleddin et al. (2013)

(continued)

Table 1 (continued)

Compound	Mechanism	Species/strain	Reinstatement	Effect on reinstatement	References
AM630	CB_2 receptor antagonist	Long-Evans rats	Cue-induced	–	Gamaleddin et al. (2012b)
AM1241	CB_2 receptor agonist	Long-Evans rats	Cue-induced	–	Gamaleddin et al. (2012b)
Rimonabant (SR141716A)	CB_1 receptor antagonist	Wistar rats	Cue-induced	↓	Diergaarde et al. (2008)
		Long-Evans rats	Cue-induced	↓	Forget et al. (2009)
		Wistar rats	Cue-induced	↓	De Vries et al. (2005)
		Sprague-Dawley rats	Cue-induced	↓	Cohen et al. (2005)
URB597	Fatty acid amide hydrolase inhibitor	Long-Evans rats	Cue-induced	↓	Forget et al. (2009)
VDM11	Anandamide transport inhibitor	Long-Evans rats	Cue-induced	↓	Gamaleddin et al. (2011)
WIN 55,212-2	$CB_{1/2}$ receptor agonist	Long-Evans rats	Cue-induced	↑	Gamaleddin et al. (2012a)
Serotonin					
M100907	$5\text{-}HT_{2A}$ receptor antagonist	Long-Evans rats	Cue-induced	↓	Fletcher et al. (2012)
Ro60-0175	$5\text{-}HT_{2C}$ receptor agonist	Long-Evans rats	Cue-induced	↓	Fletcher et al. (2012)

(continued)

Table 1 (continued)

Compound	Mechanism	Species/strain	Reinstatement	Effect on reinstatement	References
Norepinephrine					
Prazosin	Noradrenergic α1 receptor antagonist	Long-Evans rats	Cue-induced	↓	Forget et al. (2010b)
Propranolol	β-blocker	Sprague-Dawley rats	Cue-induced	↓	Chiamulera et al. (2010)
Other					
bupropion	nAChR antagonist, dopamine and norepinephrine reuptake inhibitor	Sprague-Dawley rats	Cue-induced	↑	Liu et al. (2008)
Naltrexone	Nonselective opioid antagonist	Sprague-Dawley rats	Cue-induced	↓	Liu et al. (2009)
TTA-A2	T-type calcium channel antagonist	Long-Evans rats	Cue-induced	↓	Uslaner et al. (2010)
SB334867	Hypocretin receptor-1 antagonist	C57BL/6 J mice	Cue-induced	↓	Plaza-Zabala A et al. (2013)
TCSOX229	Hypocretin receptor-2 antagonist	C57BL/6 J mice	Cue-induced	-	Plaza-Zabala A et al. (2013)
2-SORA 18	Orexin 2 receptor antagonist	Long Evans rats	Cue-induced	↓	Uslaner et al. (2014)

Table 2 Animal studies of the pharmacological targeting of various neurotransmitter systems in nicotine-induced reinstatement of nicotine seeking

Compound	Mechanism	Species/strain	Reinstatement	Effect on reinstatement	References
Acetylcholine					
Donepezil	Acetylcholinesterase inhibitor	Sprague-Dawley rats	Nicotine-induced, cues available	↓	Kimmey et al. (2012)
Galantamine	α7 and α4β2 nAChR positive allosteric modulator and acetylcholinesterase inhibitor	Sprague-Dawley rats	Nicotine-induced, cues available	↓	Hopkins et al. (2012)
varenicline	α4β2 nAChR partial agonist	Hooded Lister rats	Nicotine-induced, cues available	↓	O'Connor et al. (2010)
		Hooded Lister rats	Nicotine-induced	↓	O'Connor et al. (2010)
Glutamate					
EMQMCM	mGlu1 receptor antagonist	Wistar rats	Nicotine-induced	↓	Dravolina et al. (2007)
GABA					
Baclofen	$GABA_B$ receptor agonist	Sprague-Dawley rats	Nicotine-induced	↓	Fattore et al. (2009)
Dopamine					
Clofibrate	α-type peroxisome proliferator-activated receptor agonist	Squirrel monkeys	Nicotine-induced	↓	Panlilio et al. (2012)
L-745,870	Dopamine D_4 receptor antagonist	Long-Evans rats	Nicotine-induced	↓	Yan et al. (2013)
methOEA	α-type peroxisome proliferator-activated receptor agonist	Sprague-Dawley rats and squirrel monkeys	Nicotine-induced, cues available	↓	Mascia et al. (2011)

(continued)

Table 2 (continued)

Compound	Mechanism	Species/strain	Reinstatement	Effect on reinstatement	References
SB277011-A	Dopamine D_3 receptor antagonist	Wistar rats	Nicotine-induced, cues available for 30 min before nicotine priming	↓	Andreoli et al. (2003)
WY14643	α-type peroxisome proliferator-activated receptor agonist	Sprague-Dawley rats and squirrel monkeys	Nicotine-induced, cues available	↓	Mascia et al. (2011)
Endocannabinoid					
AM251	CB_1 receptor antagonist	Hooded Lister rats	Nicotine-induced, cues available	↓	Shoaib (2008)
AM404	CB_1 receptor antagonist	Long-Evans rats	Nicotine-induced	↓	Gamaleddin et al. (2013)
AM630	CB_2 receptor antagonist	Long-Evans rats	Nicotine-induced	–	Gamaleddin et al. (2012b)
AM1241	CB_2 receptor agonist	Long-Evans rats	Nicotine-induced	–	Gamaleddin et al. (2012b)
Rimonabant	CB_1 receptor antagonist	Long-Evans rats	Nicotine-induced	↓	Forget et al. (2009)
URB597	Fatty acid amide hydrolase inhibitor	Long-Evans rats	Nicotine-induced	↓	Forget et al. (2009)
URB597	Fatty acid amide hydrolase inhibitor	Long-Evans rats and Sprague-Dawley rats	Nicotine-induced	↓	Scherma et al. (2008)
VDM11	Anandamide transport inhibitor	Long-Evans rats	Nicotine-induced	↓	Gamaleddin et al. (2011)

(continued)

Table 2 (continued)

Compound	Mechanism	Species/strain	Reinstatement	Effect on reinstatement	References
Serotonin					
Locaserin	5-HT_{2C} receptor agonist	Sprague-Dawley rats	Nicotine-induced, cues available	↓	Higgins et al. (2012)
M100907	5-HT_{2A} receptor antagonist	Long-Evans rats	Nicotine-induced	↓	Fletcher et al. (2012)
Ro60-0175	5-HT_{2C} receptor agonist	Long-Evans rats	Nicotine-induced	↓	(Fletcher et al. 2012)
Norepinephrine					
Prazosin	Noradrenergic α1 receptor antagonist	Long-Evans rats	Nicotine-induced	↓	Forget et al. (2010b)
Other					
TTA-A2	T-type calcium channel antagonist	Long-Evans rats	Nicotine-induced, cues available during extinction and reinstatement	↓	Uslaner et al. (2010)
2-SORA 18	Orexin 2 receptor antagonist	Long Evans rats	Nicotine-induced	-	Uslaner et al. (2014)

effectiveness of these medications may be explained by the differential regulation of diverse aspects of relapse by various neurotransmitter systems, as discussed in the introduction above, which has been extensively documented for relapse to cocaine seeking. Whereas the possibility of the regulation of the different types of reinstatement by different neurocircuits has not yet been widely explored for relapse to nicotine seeking, specific smoking cessation medications may only target one aspect of relapse, resulting in decreased overall effectiveness compared with pharmacological compounds that target the full spectrum of relapse. This concept is supported by studies of "relapse" to nicotine seeking in animals, which demonstrated that the reinstatement of nicotine seeking is most robust when it is elicited by both nicotine priming and exposure to conditioned cues compared with nicotine priming or conditioned cues themselves (Feltenstein et al. 2012; O'Connor et al. 2010). With less than 5 % of all smoking cessation attempts resulting in lifelong abstinence from tobacco (Hughes et al. 2004), novel medications that attenuate relapse rates are needed. Pharmacologically targeting the neurocircuits or neurotransmitters that are involved in multiple aspects of relapse to tobacco consumption may improve smoking cessation rates. Preclinical models of relapse are greatly important in this pursuit of novel pharmacological targets in the treatment of tobacco dependence. The reinstatement procedure, an animal model widely used to assess "relapse" to drug seeking in animals, has provided important insights into the neurobiological effects of stimuli that trigger relapse in humans, most notably nicotine and its conditioned cues (Shaham et al. 2003; but see Katz and Higgins 2003). The majority of nicotine reinstatement studies have assessed nicotine reinstatement primed by nicotine or its conditioned cues separately, allowing for differentiation of the neurotransmitters that regulate these two different manipulations that induce the reinstatement of drug seeking.

Preclinical studies suggest the limited effectiveness of two widely used smoking cessation medications, varenicline and bupropion, in attenuating the cue-induced reinstatement of nicotine seeking (Liu et al. 2008; O'Connor et al. 2010, Wouda et al. 2011). These results are consistent with the clinical observations that these two FDA-approved medications are not very efficacious in attenuating tobacco smoking in humans. Furthermore, these results suggest that the pursuit of the identification of novel smoking cessation medications would be best served by developing pharmacological compounds that effectively treat the various factors that can induce the reinstatement of nicotine seeking, including nicotine and cues. nAChR subunits other than the $\alpha4\beta2$-containing nAChRs (the nAChR subtype on which varenicline acts), including $\alpha3$, $\alpha5$, and $\beta4$ subunits, may be interesting in the development of more efficacious smoking cessation medications. An increasing body of preclinical studies also suggests that exploring other neurotransmitter systems downstream from nAChRs may be lucrative in the quest for novel pharmacological targets to attenuate nicotine reinstatement.

Glutamatergic neurotransmitter systems are particularly appealing targets for the development of novel smoking cessation medications (Liechti and Markou 2008). Glutamate critically regulates the reinstatement of nicotine seeking (Bespalov et al. 2005; Gipson et al. 2013; Liechti et al. 2007) but also cocaine seeking (for reviews, see Kalivas 2004; Wise 2009), suggesting that pharmacologically targeting glutamatergic neurotransmission may be particularly promising in the identification of novel targets for smoking cessation medications. Nicotine reinstatement studies found that pharmacological compounds that decrease glutamatergic neurotransmission effectively attenuated both the cue- and nicotine-induced reinstatement of nicotine seeking (Bespalov et al. 2005; Dravolina et al. 2007; Gipson et al. 2013; Liechti et al. 2007; Pechnick et al. 2011; Ramirez-Nino et al. 2013). The promise of targeting glutamatergic compounds as smoking cessation medications is further demonstrated by the success of N-acetylcysteine in reducing cigarette consumption in smokers (Knackstedt et al. 2009).

Other than glutamatergic neurotransmission, preclinical studies have suggested that both the cue- and nicotine-induced reinstatement of nicotine seeking can be attenuated by pharmacologically targeting various other neurotransmitters (Table 1). Dopamine receptor antagonists, $GABA_B$ receptor agonists or positive allosteric modulators, endocannabinoid receptor antagonists, serotonin receptor agonists, β-blockers, and opioid receptor antagonists have been identified as potentially efficacious as pharmacological smoking cessation medications. These preclinical studies of nicotine seeking have greatly benefited from the synthesis and characterization of novel pharmacological compounds that attenuate relapse to tobacco smoking in humans. Additionally, preclinical studies on the neurobiology of cue- and nicotine-induced reinstatement provide important insights into potential future study directions for clinical trials in ongoing efforts to repurposing medications that have been approved by the FDA for other neurobiological disorders to serve as smoking cessation aids.

Acknowledgements This work was supported by Postdoctoral Fellowship 21FT-0022 from the Tobacco-Related Disease Research Program to AKS and research grant R56 DA011946 from the National Institute on Drug Abuse to AM. The authors would like to thank Ms. Janet Hightower for assistance with preparation of the figure and Mr. Michael Arends for editorial assistance.

References

Abdolahi A, Acosta G, Breslin FJ, Hemby SE, Lynch WJ (2010) Incubation of nicotine seeking is associated with enhanced protein kinase A-regulated signaling of dopamine- and cAMP-regulated phosphoprotein of 32 kDa in the insular cortex. Eur J Neurosci 31:733–741

Alberg AJ, Shopland DR, Cummings KM (2014) The 2014 Surgeon general's report: commemorating the 50th anniversary of the 1964 report of the advisory committee to the us surgeon general and updating the evidence on the health consequences of cigarette smoking. Am J Epidemiol 179:403

Andreoli M, Tessari M, Pilla M, Valerio E, Hagan JJ, Heidbreder CA (2003) Selective antagonism at dopamine D3 receptors prevents nicotine-triggered relapse to nicotine-seeking behavior. Neuropsychopharmacology 28:1272–1280

Bedi G, Preston KL, Epstein DH, Heishman SJ, Marrone GF, Shaham Y, de Wit H (2011) Incubation of cue-induced cigarette craving during abstinence in human smokers. Biol Psychiatry 69:708–711

Beltramo M, Stella N, Calignano A, Lin SY, Makriyannis A, Piomelli D (1997) Functional role of high-affinity Anandamide transport, as revealed by selective inhibition. Science 277:1094–1097
Berlin I, Said S, Spreux-Varoquaux O, Launay JM, Olivares R, Millet V, Lecrubier Y, Puech AJ (1995) A reversible monoamine oxidase A inhibitor (moclobemide) facilitates smoking cessation and abstinence in heavy, dependent smokers. Clin Pharmacol Ther 58:444–452
Bespalov AY, Dravolina OA, Sukhanov I, Zakharova E, Blokhina E, Zvartau E, Danysz W, van Heeke G, Markou A (2005) Metabotropic glutamate receptor (mGluR5) antagonist MPEP attenuated cue- and schedule-induced reinstatement of nicotine self-administration behavior in rats. Neuropharmacology 49:167–178
Biggs JT, Spiker DG, Petit JM, Ziegler VE (1977) Tricyclic antidepressant overdose: incidence of symptoms. JAMA 238:135–138
Bormann J (1986) Electrophysiology of single GABA receptor chloride channels. Clin Neuropharmacol 9:386–388
Bruijnzeel AW, Prado M, Isaac S (2009) Corticotropin-releasing factor-1 receptor activation mediates nicotine withdrawal-induced deficit in brain reward function and stress-induced relapse. Biol Psychiatry 66:110–117
Cahill K, Ussher MH (2011) Cannabinoid type 1 receptor antagonists for smoking cessation. Cochrane Database Syst Rev 3:CD005353
Centers for Disease Control and Prevention (2000) Cigarette smoking among adults - United States, 1998. Morb Mortal Wkly Rep 49:881–884
Centers for Disease Control and Prevention (2010) Quitting smoking among adults—United States 2001–2010. Morb Mortal Wkly Rep 60:1513–1519
Chiamulera C, Tedesco V, Zangrandi L, Giuliano C, Fumagalli G (2010) Propranolol transiently inhibits reinstatement of nicotine-seeking behaviour in rats. J Psychopharmacol 24:389–395
Clinicaltrials.gov (2007) Effects of AFQ056 and nicotine in reducing cigarette smoking. http://clinicaltrials.gov/ct2/show/NCT00414752?term=Effects+of+AFQ056+and+Nicotine+in+Reducing+Cigarette+Smoking&rank=1. Accessed 23 Mar 2014
Coe JW, Brooks PR, Vetelino MG, Wirtz MC, Arnold EP, Huang J, Sands SB, Davis TI, Lebel LA, Fox CB, Shrikhande A, Heym JH, Schaeffer E, Rollema H, Lu Y, Mansbach RS, Chambers LK, Rovetti CC, Schulz DW, Tingley FD 3rd, O'Neill BT (2005) Varenicline: an alpha4beta2 nicotinic receptor partial agonist for smoking cessation. J Med Chem 48:3474–3477
Cohen C, Perrault G, Griebel G, Soubrie P (2005) Nicotine-associated cues maintain nicotine-seeking behavior in rats several weeks after nicotine withdrawal: reversal by the cannabinoid (CB1) receptor antagonist, rimonabant (SR141716). Neuropsychopharmacology 30:145–155
Contreras M, Billeke P, Vicencio S, Madrid C, Perdomo G, Gonzalez M, Torrealba F (2012) A role for the insular cortex in long-term memory for context-evoked drug craving in rats. Neuropsychopharmacology 37:2101–2108
Contreras M, Ceric F, Torrealba F (2007) Inactivation of the interoceptive insula disrupts drug craving and malaise induced by lithium. Science 318:655–658
Cousins MS, Stamat HM, de Wit H (2001) Effects of a single dose of baclofen on self-reported subjective effects and tobacco smoking. Nicotine Tob Res 3:123–129
Cravatt BF, Giang DK, Mayfield SP, Boger DL, Lerner RA, Gilula NB (1996) Molecular characterization of an enzyme that degrades neuromodulatory fatty-acid amides. Nature 384:83–87
Cryan JF, Bruijnzeel AW, Skjei KL, Markou A (2003) Bupropion enhances brain reward function and reverses the affective and somatic aspects of nicotine withdrawal in the rat. Psychopharmacology 168:347–358
De Biasi M, Salas R (2008) Influence of neuronal nicotinic receptors over nicotine addiction and withdrawal. Exp Biol Med (Maywood) 233:917–929
De Vries TJ, de Vries W, Janssen MC, Schoffelmeer AN (2005) Suppression of conditioned nicotine and sucrose seeking by the cannabinoid-1 receptor antagonist SR141716A. Behav Brain Res 161:164–168

Diergaarde L, de Vries W, Raaso H, Schoffelmeer AN, De Vries TJ (2008) Contextual renewal of nicotine seeking in rats and its suppression by the cannabinoid-1 receptor antagonist Rimonabant (SR141716A). Neuropharmacology 55:712–716

Doherty K, Kinnunen T, Militello FS, Garvey AJ (1995) Urges to smoke during the first month of abstinence: relationship to relapse and predictors. Psychopharmacology 119:171–178

Dravolina OA, Zakharova ES, Shekunova EV, Zvartau EE, Danysz W, Bespalov AY (2007) mGlu1 receptor blockade attenuates cue- and nicotine-induced reinstatement of extinguished nicotine self-administration behavior in rats. Neuropharmacology 52:263–269

Edwards NB, Murphy JK, Downs AD, Ackerman BJ, Rosenthal TL (1989) Doxepin as an adjunct to smoking cessation: a double-blind pilot study. Am J Psychiatry 146:373–376

Epstein AM, King AC (2004) Naltrexone attenuates acute cigarette smoking behavior. Pharmacol Biochem Behav 77:29–37

Everitt BJ, Belin D, Economidou D, Pelloux Y, Dalley JW, Robbins TW (2008) Review neural mechanisms underlying the vulnerability to develop compulsive drug-seeking habits and addiction. Philos Trans R Soc Lond B Biol Sci 363:3125–3135

Falvella FS, Galvan A, Frullanti E, Spinola M, Calabro E, Carbone A, Incarbone M, Santambrogio L, Pastorino U, Dragani TA (2009) Transcription deregulation at the 15q25 locus in association with lung adenocarcinoma risk. Clin Cancer Res 15:1837–1842

Fattore L, Spano MS, Cossu G, Scherma M, Fratta W, Fadda P (2009) Baclofen prevents drug-induced reinstatement of extinguished nicotine-seeking behaviour and nicotine place preference in rodents. Eur Neuropsychopharmacol 19:487–498

Fegley D, Gaetani S, Duranti A, Tontini A, Mor M, Tarzia G, Piomelli D (2005) Characterization of the fatty acid amide hydrolase inhibitor cyclohexyl carbamic acid 3'-carbamoyl-biphenyl-3-yl ester (URB597): effects on anandamide and oleoylethanolamide deactivation. J Pharmacol Exp Ther 313:352–358

Feltenstein MW, Ghee SM, See RE (2012) Nicotine self-administration and reinstatement of nicotine-seeking in male and female rats. Drug Alcohol Depend 121:240–246

Fletcher PJ, Le AD, Higgins GA (2008) Serotonin receptors as potential targets for modulation of nicotine use and dependence. Prog Brain Res 172:361–383

Fletcher PJ, Rizos Z, Noble K, Soko AD, Silenieks LB, Le AD, Higgins GA (2012) Effects of the 5-HT2C receptor agonist Ro60-0175 and the 5-HT2A receptor antagonist M100907 on nicotine self-administration and reinstatement. Neuropharmacology 62:2288–2298

Forget B, Coen KM, Le Foll B (2009) Inhibition of fatty acid amide hydrolase reduces reinstatement of nicotine seeking but not break point for nicotine self-administration–comparison with CB(1) receptor blockade. Psychopharmacology 205:613–624

Forget B, Pushparaj A, Le Foll B (2010a) Granular insular cortex inactivation as a novel therapeutic strategy for nicotine addiction. Biol Psychiatry 68:265–271

Forget B, Wertheim C, Mascia P, Pushparaj A, Goldberg SR, Le Foll B (2010b) Noradrenergic alpha1 receptors as a novel target for the treatment of nicotine addiction. Neuropsychopharmacology 35:1751–1760

Fowler CD, Lu Q, Johnson PM, Marks MJ, Kenny PJ (2011) Habenular alpha5 nicotinic receptor subunit signalling controls nicotine intake. Nature 471:597–601

Fowler CD, Tuesta L, Kenny PJ (2013) Role of alpha5* nicotinic acetylcholine receptors in the effects of acute and chronic nicotine treatment on brain reward function in mice. Psychopharmacology 229:503–513

Frahm S, Slimak MA, Ferrarese L, Santos-Torres J, Antolin-Fontes B, Auer S, Filkin S, Pons S, Fontaine JF, Tsetlin V, Maskos U, Ibanez-Tallon I (2011) Aversion to nicotine is regulated by the balanced activity of beta4 and alpha5 nicotinic receptor subunits in the medial habenula. Neuron 70:522–535

French ED (1997) delta9-Tetrahydrocannabinol excites rat VTA dopamine neurons through activation of cannabinoid CB1 but not opioid receptors. Neurosci Lett 226:159–162

French ED, Dillon K, Wu X (1997) Cannabinoids excite dopamine neurons in the ventral tegmentum and substantia nigra. NeuroReport 8:649–652

Fuchs RA, Branham RK, See RE (2006) Different neural substrates mediate cocaine seeking after abstinence versus extinction training: a critical role for the dorsolateral caudate-putamen. J Neurosci 26:3584–3588

Fucile S, Barabino B, Palma E, Grassi F, Limatola C, Mileo AM, Alema S, Ballivet M, Eusebi F (1997) Alpha 5 subunit forms functional alpha 3 beta 4 alpha 5 nAChRs in transfected human cells. NeuroReport 8:2433–2436

Fung YK, Richard LA (1994) Behavioural consequences of cocaine withdrawal in rats. J Pharm Pharmacol 46:150–152

Gabriele A, See RE (2011) Lesions and reversible inactivation of the dorsolateral caudate-putamen impair cocaine-primed reinstatement to cocaine-seeking in rats. Brain Res 1417:27–35

Gamaleddin I, Guranda M, Goldberg SR, Le Foll B (2011) The selective Anandamide transport inhibitor VDM11 attenuates reinstatement of nicotine seeking behaviour, but does not affect nicotine intake. Br J Pharmacol 164:1652–1660

Gamaleddin I, Guranda M, Scherma M, Fratta W, Makriyannis A, Vadivel SK, Goldberg SR, Le Foll B (2013) AM404 attenuates reinstatement of nicotine seeking induced by nicotine-associated cues and nicotine priming but does not affect nicotine- and food-taking. J Psychopharmacol 27:564–571

Gamaleddin I, Wertheim C, Zhu AZ, Coen KM, Vemuri K, Makryannis A, Goldberg SR, Le Foll B (2012a) Cannabinoid receptor stimulation increases motivation for nicotine and nicotine seeking. Addict Biol 17:47–61

Gamaleddin I, Zvonok A, Makriyannis A, Goldberg SR, Le Foll B (2012b) Effects of a selective cannabinoid CB2 agonist and antagonist on intravenous nicotine self administration and reinstatement of nicotine seeking. PLoS ONE 7:e29900

Gardner EL (2005) Endocannabinoid signaling system and brain reward: emphasis on dopamine. Pharmacol Biochem Behav 81:263–284

Gardner EL, Vorel SR (1998) Cannabinoid transmission and reward-related events. Neurobiol Dis 5:502–533

Gass JT, Olive MF (2008) Glutamatergic substrates of drug addiction and alcoholism. Biochem Pharmacol 75:218–265

Gerasimov MR, Franceschi M, Volkow ND, Rice O, Schiffer WK, Dewey SL (2000) Synergistic interactions between nicotine and cocaine or methylphenidate depend on the dose of dopamine transporter inhibitor. Synapse 38:432–437

Geyer MA, Markou A (1995) Animal models of psychiatric disorders. In: Bloom FE, Kupfer DJ (eds) Psychopharmacology: the fourth generation of progress. Raven Press, New York, pp 787–798

Giang DK, Cravatt BF (1997) Molecular characterization of human and mouse fatty acid amide hydrolases. Proc Natl Acad Sci USA 94:2238–2242

Gipson CD, Reissner KJ, Kupchik YM, Smith AC, Stankeviciute N, Hensley-Simon ME, Kalivas PW (2013) Reinstatement of nicotine seeking is mediated by glutamatergic plasticity. Proc Natl Acad Sci USA 110:9124–9129

Gonzales D, Rennard SI, Nides M, Oncken C, Azoulay S, Billing CB, Watsky EJ, Gong J, Williams KE, Reeves KR (2006) Varenicline, an alpha4beta2 nicotinic acetylcholine receptor partial agonist, versus sustained-release bupropion and placebo for smoking cessation: a randomized controlled trial. JAMA 296:47–55

Guery S, Floersheim P, Kaupmann K, Froestl W (2007) Syntheses and optimization of new GS39783 analogues as positive allosteric modulators of GABA B receptors. Bioorg Med Chem Lett 17:6206–6211

Hall SM, Reus VI, Munoz RF, Sees KL, Humfleet G, Hartz DT, Frederick S, Triffleman E (1998) Nortriptyline and cognitive-behavioral therapy in the treatment of cigarette smoking. Arch Gen Psychiatry 55:683–690

Higgins GA, Silenieks LB, Rossmann A, Rizos Z, Noble K, Soko AD, Fletcher PJ (2012) The 5-HT2C receptor agonist lorcaserin reduces nicotine self-administration, discrimination, and reinstatement: relationship to feeding behavior and impulse control. Neuropsychopharmacology 37:1177–1191

Hollander JA, Lu Q, Cameron MD, Kamenecka TM, Kenny PJ (2008) Insular hypocretin transmission regulates nicotine reward. Proc Natl Acad Sci USA 105:19480–19485

Hopkins TJ, Rupprecht LE, Hayes MR, Blendy JA, Schmidt HD (2012) Galantamine, an acetylcholinesterase inhibitor and positive allosteric modulator of nicotinic acetylcholine receptors, attenuates nicotine taking and seeking in rats. Neuropsychopharmacology 37:2310–2321

Howlett AC (2002) The cannabinoid receptors. Prostaglandins Other Lipid Mediat 68–69:619–631

Howlett AC, Breivogel CS, Childers SR, Deadwyler SA, Hampson RE, Porrino LJ (2004) Cannabinoid physiology and pharmacology: 30 years of progress. Neuropharmacology 47:345–358

Hughes JR, Keely J, Naud S (2004) Shape of the relapse curve and long-term abstinence among untreated smokers. Addiction 99:29–38

Hughes JR, Shiffman S, Callas P, Zhang J (2003) A meta-analysis of the efficacy of over-the-counter nicotine replacement. Tob Control 12:21–27

Hung RJ, McKay JD, Gaborieau V, Boffetta P, Hashibe M, Zaridze D, Mukeria A, Szeszenia-Dabrowska N, Lissowska J, Rudnai P, Fabianova E, Mates D, Bencko V, Foretova L, Janout V, Chen C, Goodman G, Field JK, Liloglou T, Xinarianos G, Cassidy A, McLaughlin J, Liu G, Narod S, Krokan HE, Skorpen F, Elvestad MB, Hveem K, Vatten L, Linseisen J, Clavel-Chapelon F, Vineis P, Bueno-de-Mesquita HB, Lund E, Martinez C, Bingham S, Rasmuson T, Hainaut P, Riboli E, Ahrens W, Benhamou S, Lagiou P, Trichopoulos D, Holcatova I, Merletti F, Kjaerheim K, Agudo A, Macfarlane G, Talamini R, Simonato L, Lowry R, Conway DI, Znaor A, Healy C, Zelenika D, Boland A, Delepine M, Foglio M, Lechner D, Matsuda F, Blanche H, Gut I, Heath S, Lathrop M, Brennan P (2008) A susceptibility locus for lung cancer maps to nicotinic acetylcholine receptor subunit genes on 15q25. Nature 452:633–637

Hurt RD, Sachs DP, Glover ED, Offord KP, Johnston JA, Dale LC, Khayrallah MA, Schroeder DR, Glover PN, Sullivan CR, Croghan IT, Sullivan PM (1997) A comparison of sustained-release bupropion and placebo for smoking cessation. N Engl J Med 337:1195–1202

Ito R, Dalley JW, Robbins TW, Everitt BJ (2002) Dopamine release in the dorsal striatum during cocaine-seeking behavior under the control of a drug-associated cue. J Neurosci 22:6247–6253

Jackson KJ, Martin BR, Changeux JP, Damaj MI (2008) Differential role of nicotinic acetylcholine receptor subunits in physical and affective nicotine withdrawal signs. J Pharmacol Exp Ther 325:302–312

Jorenby DE, Hays JT, Rigotti NA, Azoulay S, Watsky EJ, Williams KE, Billing CB, Gong J, Reeves KR (2006) Efficacy of varenicline, an alpha4beta2 nicotinic acetylcholine receptor partial agonist, vs placebo or sustained-release bupropion for smoking cessation: a randomized controlled trial. JAMA 296:56–63

Kalivas PW (2004) Glutamate systems in cocaine addiction. Curr Opin Pharmacol 4:23–29

Kalivas PW, McFarland K (2003) Brain circuitry and the reinstatement of cocaine-seeking behavior. Psychopharmacology 168:44–56

Katz JL, Higgins ST (2003) The validity of the reinstatement model of craving and relapse to drug use. Psychopharmacology 168:21–30

Khaled MA, Farid Araki K, Li B, Coen KM, Marinelli PW, Varga J, Gaal J, Le Foll B (2010) The selective dopamine D3 receptor antagonist SB 277011-A, but not the partial agonist BP 897, blocks cue-induced reinstatement of nicotine-seeking. Int J Neuropsychopharmacol 13:181–190

Kimmey BA, Rupprecht LE, Hayes MR, Schmidt HD (2012) Donepezil, an acetylcholinesterase inhibitor, attenuates nicotine self-administration and reinstatement of nicotine seeking in rats. Addict Biol 19:539–51

King AC (2002) Role of naltrexone in initial smoking cessation: preliminary findings. Alcohol Clin Exp Res 26:1942–1944

King AC, Cao D, O'Malley SS, Kranzler HR, Cai X, deWit H, Matthews AK, Stachoviak RJ (2012) Effects of naltrexone on smoking cessation outcomes and weight gain in nicotine-dependent men and women. J Clin Psychopharmacol 32:630–636

King AC, Cao D, Zhang L, O'Malley SS (2013) Naltrexone reduction of long-term smoking cessation weight gain in women but not men: a randomized controlled trial. Biol Psychiatry 73:924–930

King AC, Meyer PJ (2000) Naltrexone alteration of acute smoking response in nicotine-dependent subjects. Pharmacol Biochem Behav 66:563–572

Klitenick MA, DeWitte P, Kalivas PW (1992) Regulation of somatodendritic dopamine release in the ventral tegmental area by opioids and GABA: an in vivo microdialysis study. J Neurosci 12:2623–2632

Knackstedt LA, LaRowe S, Mardikian P, Malcolm R, Upadhyaya H, Hedden S, Markou A, Kalivas PW (2009) The role of cystine-glutamate exchange in nicotine dependence in rats and humans. Biol Psychiatry 65:841–845

Kupchik YM, Moussawi K, Tang XC, Wang X, Kalivas BC, Kolokithas R, Ogburn KB, Kalivas PW (2012) The effect of N-acetylcysteine in the nucleus accumbens on neurotransmission and relapse to cocaine. Biol Psychiatry 71:978–986

Laviolette SR, van der Kooy D (2004) The neurobiology of nicotine addiction: bridging the gap from molecules to behaviour. Nat Rev Neurosci 5:55–65

Le Foll B, Chakraborty-Chatterjee M, Lev-Ran S, Barnes C, Pushparaj A, Gamaleddin I, Yan Y, Khaled M, Goldberg SR (2012) Varenicline decreases nicotine self-administration and cue-induced reinstatement of nicotine-seeking behaviour in rats when a long pretreatment time is used. Int J Neuropsychopharmacol 15:1265–1274

Li S, Li Z, Pei L, Le AD, Liu F (2012) The alpha7nACh-NMDA receptor complex is involved in cue-induced reinstatement of nicotine seeking. J Exp Med 209:2141–2147

Li X, Semenova S, D'Souza MS, Stoker AK, Markou A (2014) Involvement of glutamatergic and GABAergic systems in nicotine dependence: Implications for novel pharmacotherapies for smoking cessation. Neuropharmacology Part B 76:554–565

Liechti ME, Lhuillier L, Kaupmann K, Markou A (2007) Metabotropic glutamate 2/3 receptors in the ventral tegmental area and the nucleus accumbens shell are involved in behaviors relating to nicotine dependence. J Neurosci 27:9077–9085

Liechti ME, Markou A (2008) Role of the glutamatergic system in nicotine dependence : implications for the discovery and development of new pharmacological smoking cessation therapies. CNS Drugs 22:705–724

Liu X (2014) Effects of blockade of alpha4beta2 and alpha7 nicotinic acetylcholine receptors on cue-induced reinstatement of nicotine-seeking behaviour in rats. Int J Neuropsychopharmacol 17:105–116

Liu X, Caggiula AR, Palmatier MI, Donny EC, Sved AF (2008) Cue-induced reinstatement of nicotine-seeking behavior in rats: effect of bupropion, persistence over repeated tests, and its dependence on training dose. Psychopharmacology 196:365–375

Liu X, Caggiula AR, Yee SK, Nobuta H, Sved AF, Pechnick RN, Poland RE (2007) Mecamylamine attenuates cue-induced reinstatement of nicotine-seeking behavior in rats. Neuropsychopharmacology 32:710–718

Liu X, Jernigen C, Gharib M, Booth S, Caggiula AR, Sved AF (2010) Effects of dopamine antagonists on drug cue-induced reinstatement of nicotine-seeking behavior in rats. Behav Pharmacol 21:153–160

Liu X, Palmatier MI, Caggiula AR, Sved AF, Donny EC, Gharib M, Booth S (2009) Naltrexone attenuation of conditioned but not primary reinforcement of nicotine in rats. Psychopharmacology 202:589–598

Mameli-Engvall M, Evrard A, Pons S, Maskos U, Svensson TH, Changeux JP, Faure P (2006) Hierarchical control of dopamine neuron-firing patterns by nicotinic receptors. Neuron 50:911–921

Mansvelder HD, McGehee DS (2000) Long-term potentiation of excitatory inputs to brain reward areas by nicotine. Neuron 27:349–357

Markou A (2007) Metabotropic glutamate receptor antagonists: novel therapeutics for nicotine dependence and depression? Biol Psychiatry 61:17–22

Mascia P, Pistis M, Justinova Z, Panlilio LV, Luchicchi A, Lecca S, Scherma M, Fratta W, Fadda P, Barnes C, Redhi GH, Yasar S, Le Foll B, Tanda G, Piomelli D, Goldberg SR (2011) Blockade of nicotine reward and reinstatement by activation of alpha-type peroxisome proliferator-activated receptors. Biol Psychiatry 69:633–641

Maskos U, Molles BE, Pons S, Besson M, Guiard BP, Guilloux JP, Evrard A, Cazala P, Cormier A, Mameli-Engvall M, Dufour N, Cloez-Tayarani I, Bemelmans AP, Mallet J, Gardier AM, David V, Faure P, Granon S, Changeux JP (2005) Nicotine reinforcement and cognition restored by targeted expression of nicotinic receptors. Nature 436:103–107

Mihalak KB, Carroll FI, Luetje CW (2006) Varenicline is a partial agonist at alpha4beta2 and a full agonist at alpha7 neuronal nicotinic receptors. Mol Pharmacol 70:801–805

Moreira FA, Crippa JA (2009) The psychiatric side-effects of rimonabant. Rev Bras Psiquiatr 31:145–153

Muskens JB, Schellekens AF, de Leeuw FE, Tendolkar I, Hepark S (2012) Damage in the dorsal striatum alleviates addictive behavior. Gen Hosp Psychiatry 34: 702 e9-702 e11

Naqvi NH, Bechara A (2009) The hidden island of addiction: the insula. Trends Neurosci 32:56–67

Naqvi NH, Rudrauf D, Damasio H, Bechara A (2007) Damage to the insula disrupts addiction to cigarette smoking. Science 315:531–534

Ng M, Freeman MK, Fleming TD, Robinson M, Dwyer-Lindgren L, Thomson B, Wollum A, Sanman E, Wulf S, Lopez AD, Murray CJ, Gakidou E (2014) Smoking prevalence and cigarette consumption in 187 countries, 1980-2012. JAMA 311:183–192

O'Connor EC, Parker D, Rollema H, Mead AN (2010) The alpha4beta2 nicotinic acetylcholine-receptor partial agonist varenicline inhibits both nicotine self-administration following repeated dosing and reinstatement of nicotine seeking in rats. Psychopharmacology 208:365–376

Panlilio LV, Justinova Z, Mascia P, Pistis M, Luchicchi A, Lecca S, Barnes C, Redhi GH, Adair J, Heishman SJ, Yasar S, Aliczki M, Haller J, Goldberg SR (2012) Novel use of a lipid-lowering fibrate medication to prevent nicotine reward and relapse: preclinical findings. Neuropsychopharmacology 37:1838–1847

Paterson NE, Balfour DJ, Markou A (2007) Chronic bupropion attenuated the anhedonic component of nicotine withdrawal in rats via inhibition of dopamine reuptake in the nucleus accumbens shell. Eur J Neurosci 25:3099–3108

Paterson NE, Froestl W, Markou A (2005) Repeated administration of the GABAB receptor agonist CGP44532 decreased nicotine self-administration, and acute administration decreased cue-induced reinstatement of nicotine-seeking in rats. Neuropsychopharmacology 30:119–128

Pechnick RN, Manalo CM, Lacayo LM, Vit JP, Bholat Y, Spivak I, Reyes KC, Farrokhi C (2011) Acamprosate attenuates cue-induced reinstatement of nicotine-seeking behavior in rats. Behav Pharmacol 22:222–227

Picciotto MR (1998) Common aspects of the action of nicotine and other drugs of abuse. Drug Alcohol Depend 51:165–172

Plaza-Zabala A, Flores Á, Martín-García E, Saravia R, Maldonado R, Berrendero F (2013) A role for hypocretin/orexin receptor-1 in cue-induced reinstatement of nicotine-seeking behavior. Neuropsychopharmacology 38:1724–36

Prochazka AV, Weaver MJ, Keller RT, Fryer GE, Licari PA, Lofaso D (1998) A randomized trial of nortriptyline for smoking cessation. Arch Intern Med 158:2035–2039

Pushparaj A, Hamani C, Yu W, Shin DS, Kang B, Nobrega JN, Le Foll B (2013) Electrical stimulation of the insular region attenuates nicotine-taking and nicotine-seeking behaviors. Neuropsychopharmacology 38:690–698

Ramirez-Nino AM, D'Souza MS, Markou A (2013) N-acetylcysteine decreased nicotine self-administration and cue-induced reinstatement of nicotine seeking in rats: comparison with the effects of N-acetylcysteine on food responding and food seeking. Psychopharmacology 225:473–482

Roose SP, Dalack GW, Glassman AH, Woodring S, Walsh BT, Giardina EG (1991) Is doxepin a safer tricyclic for the heart? J Clin Psychiatry 52:338–341

Saccone NL, Culverhouse RC, Schwantes-An TH, Cannon DS, Chen X, Cichon S, Giegling I, Han S, Han Y, Keskitalo-Vuokko K, Kong X, Landi MT, Ma JZ, Short SE, Stephens SH, Stevens VL, Sun L, Wang Y, Wenzlaff AS, Aggen SH, Breslau N, Broderick P, Chatterjee N, Chen J, Heath AC, Heliovaara M, Hoft NR, Hunter DJ, Jensen MK, Martin NG, Montgomery

GW, Niu T, Payne TJ, Peltonen L, Pergadia ML, Rice JP, Sherva R, Spitz MR, Sun J, Wang JC, Weiss RB, Wheeler W, Witt SH, Yang BZ, Caporaso NE, Ehringer MA, Eisen T, Gapstur SM, Gelernter J, Houlston R, Kaprio J, Kendler KS, Kraft P, Leppert MF, Li MD, Madden PA, Nothen MM, Pillai S, Rietschel M, Rujescu D, Schwartz A, Amos CI, Bierut LJ (2010) Multiple independent loci at chromosome 15q25.1 affect smoking quantity: a meta-analysis and comparison with lung cancer and COPD. PLoS Genet 6:e1001053

Saccone SF, Hinrichs AL, Saccone NL, Chase GA, Konvicka K, Madden PA, Breslau N, Johnson EO, Hatsukami D, Pomerleau O, Swan GE, Goate AM, Rutter J, Bertelsen S, Fox L, Fugman D, Martin NG, Montgomery GW, Wang JC, Ballinger DG, Rice JP, Bierut LJ (2007) Cholinergic nicotinic receptor genes implicated in a nicotine dependence association study targeting 348 candidate genes with 3713 SNPs. Hum Mol Genet 16:36–49

Salas R, Main A, Gangitano D, De Biasi M (2007) Decreased withdrawal symptoms but normal tolerance to nicotine in mice null for the alpha7 nicotinic acetylcholine receptor subunit. Neuropharmacology 53:863–869

Scherma M, Panlilio LV, Fadda P, Fattore L, Gamaleddin I, Le Foll B, Justinova Z, Mikics E, Haller J, Medalie J, Stroik J, Barnes C, Yasar S, Tanda G, Piomelli D, Fratta W, Goldberg SR (2008) Inhibition of anandamide hydrolysis by cyclohexyl carbamic acid 3'-carbamoyl-3-yl ester (URB597) reverses abuse-related behavioral and neurochemical effects of nicotine in rats. J Pharmacol Exp Ther 327:482–490

Schlicker E, Kathmann M (2001) Modulation of transmitter release via presynaptic cannabinoid receptors. Trends Pharmacol Sci 22:565–572

Scott D, Hiroi N (2011) Deconstructing craving: dissociable cortical control of cue reactivity in nicotine addiction. Biol Psychiatry 69:1052–1059

See RE, Fuchs RA, Ledford CC, McLaughlin J (2003) Drug addiction, relapse, and the amygdala. Ann N Y Acad Sci 985:294–307

Shaham Y, Shalev U, Lu L, De Wit H, Stewart J (2003) The reinstatement model of drug relapse: history, methodology and major findings. Psychopharmacology 168:3–20

Shoaib M (2008) The cannabinoid antagonist AM251 attenuates nicotine self-administration and nicotine-seeking behaviour in rats. Neuropharmacology 54:438–44

Slemmer JE, Martin BR, Damaj MI (2000) Bupropion is a nicotinic antagonist. J Pharmacol Exp Ther 295:321–327

Stabl M, Biziere K, Schmid-Burgk W, Amrein R (1989) Review of comparative clinical trials. Moclobemide vs tricyclic antidepressants and vs placebo in depressive states. J Neural Transm 28:77–89

Stoker AK, Olivier B, Markou A (2012) Role of alpha7- and beta4-containing nicotinic acetylcholine receptors in the affective and somatic aspects of nicotine withdrawal: studies in knockout mice. Behav Genet 42:423–436

Stolerman IP, Jarvis MJ (1995) The scientific case that nicotine is addictive. Psychopharmacology 117: 2–10:14–20

Tanda G, Pontieri FE, Di Chiara G (1997) Cannabinoid and heroin activation of mesolimbic dopamine transmission by a common mu1 opioid receptor mechanism. Science 276:2048–2050

Thompson GH, Hunter DA (1998) Nicotine replacement therapy. Ann Pharmacother 32:1067–1075

Uslaner JM, Smith SM, Huszar SL, Pachmerhiwala R, Hinchliffe RM, Vardigan JD, Nguyen SJ, Surles NO, Yao L, Barrow JC, Uebele VN, Renger JJ, Clark J, Hutson PH (2012) T-type calcium channel antagonism produces antipsychotic-like effects and reduces stimulant-induced glutamate release in the nucleus accumbens of rats. Neuropharmacology 62:1413–1421

Uslaner JM, Vardigan JD, Drott JM, Uebele VN, Renger JJ, Lee A, Li Z, Le AD, Hutson PH (2010) T-type calcium channel antagonism decreases motivation for nicotine and blocks nicotine- and cue-induced reinstatement for a response previously reinforced with nicotine. Biol Psychiatry 68:712–718

Uslaner JM, Winrow CJ, Gotter AL, Roecker AJ, Coleman PJ, Hutson PH, Le AD, Renger JJ (2014) Selective orexin 2 receptor antagonism blocks cue-induced reinstatement, but not nicotine self-administration or nicotine-induced reinstatement. Behav Brain Res 269:61–5

Vlachou S, Guery S, Froestl W, Banerjee D, Benedict J, Finn MG, Markou A (2011) Repeated administration of the GABAB receptor positive modulator BHF177 decreased nicotine self-administration, and acute administration decreased cue-induced reinstatement of nicotine seeking in rats. Psychopharmacology 215:117–128

Vlachou S, Markou A (2010) GABAB receptors in reward processes. Adv Pharmacol 58:315–371

Volkow ND, Wang GJ, Telang F, Fowler JS, Logan J, Childress AR, Jayne M, Ma Y, Wong C (2006) Cocaine cues and dopamine in dorsal striatum: mechanism of craving in cocaine addiction. J Neurosci 26:6583–6588

Wang JC, Cruchaga C, Saccone NL, Bertelsen S, Liu P, Budde JP, Duan W, Fox L, Grucza RA, Kern J, Mayo K, Reyes O, Rice J, Saccone SF, Spiegel N, Steinbach JH, Stitzel JA, Anderson MW, You M, Stevens VL, Bierut LJ, Goate AM (2009) Risk for nicotine dependence and lung cancer is conferred by mRNA expression levels and amino acid change in CHRNA5. Hum Mol Genet 18:3125–3135

West R, Baker CL, Cappelleri JC, Bushmakin AG (2008) Effect of varenicline and bupropion SR on craving, nicotine withdrawal symptoms, and rewarding effects of smoking during a quit attempt. Psychopharmacology 197:371–377

Wise RA (2009) Ventral tegmental glutamate: a role in stress-, cue-, and cocaine-induced reinstatement of cocaine-seeking. Neuropharmacology 56:174–176

Wouda JA, Riga D, De Vries W, Stegeman M, van Mourik Y, Schetters D, Schoffelmeer AN, Pattij T, De Vries TJ (2011) Varenicline attenuates cue-induced relapse to alcohol, but not nicotine seeking, while reducing inhibitory response control. Psychopharmacology 216:267–277

Yan Y, Pushparaj A, Le Strat Y, Gamaleddin I, Barnes C, Justinova Z, Goldberg SR, Le Foll B (2013) Blockade of dopamine d4 receptors attenuates reinstatement of extinguished nicotine-seeking behavior in rats. Neuropsychopharmacology 37:685–696

Psychiatric Disorders as Vulnerability Factors for Nicotine Addiction: What Have We Learned from Animal Models?

Bernard Le Foll, Enoch Ng, Patricia Di Ciano and José M. Trigo

Abstract Epidemiological studies indicate a high prevalence of tobacco smoking in subjects with psychiatric disorders. Notably, there is a high prevalence of smoking among those with dependence to other substances, schizophrenia, mood, or anxiety disorders. It has been difficult to understand how these phenomena interact with clinical populations as it is unclear what preceded what in most of the studies. These comorbidities may be best understood by using experimental approaches in well-controlled conditions. Notably, animal models represent advantageous approaches as the parameters under study can be controlled perfectly. This review will focus on evidence collected so far exploring how behavioral effects of nicotine are modified in animal models of psychiatric conditions. Notably, we will focus on behavioral responses induced by nicotine that are relevant for its addictive potential. Despite the clinical relevance and frequency of the comorbidity between psychiatric issues and tobacco smoking, very few studies have been done to explore this issue in animals. The available data suggest that the behavioral and reinforcing effects of nicotine are enhanced in animal models of these comorbidities, although much more experimental work would be required to provide certainty in this domain.

B. Le Foll (✉) · P. Di Ciano · J.M. Trigo
Translational Addiction Research Laboratory, Centre for Addiction and Mental Health, 33 Russell Street, Toronto, ON M5S 2S1, Canada
e-mail: bernard.lefoll@camh.ca

B. Le Foll
Ambulatory Care and Structured Treatment Program, Centre for Addiction and Mental Health, 33 Russell Street, Toronto, ON M5S 2S1, Canada

B. Le Foll
Departments of Family and Community Medicine, Pharmacology, Psychiatry, Institute of Medical Sciences, University of Toronto, Toronto, Canada

E. Ng
Lunenfeld Tanenbaum Research Institute, Mount Sinai Hospital, 600 University Avenue, Toronto, ON M5G 1X5, Canada

B. Le Foll
Centre for Addiction and Mental Health, 33 Russell Street, Toronto, ON M5V 2W8, Canada

D.J.K. Balfour and M.R. Munafò (eds.), *The Neuropharmacology of Nicotine Dependence*, Current Topics in Behavioral Neurosciences 24,
DOI 10.1007/978-3-319-13482-6_6

Keywords Nicotine · Schizophrenia · Mood disorders · Addiction · Anxiety

Contents

1 Introduction

The high prevalence of comorbid substance use disorders (SUDs) and mental illness (MI) has been well established in both clinical- and population-based studies (Brooner et al. 1997; Hien et al. 1997; Regier et al. 1990; Swendsen et al. 2010). The Epidemiological Catchment Area (ECA) study reported that among individuals with MI, 29 % had a comorbid SUD (compared with 13 % of individuals without a MI). Findings from the National Comorbidity Survey report a significantly higher prevalence of SUDs among individuals with MI than in the general population; among individuals with any MI, 51 % were reported as having a comorbid SUD, and the odds ratio for a SUD among individuals with a MI was 2.4 (compared to those without a MI) (Kessler et al. 1996a, b).

A recent study has explored the issue of transition from drug use to SUD using the National Epidemiological Survey of Alcohol and Related Conditions (NESARC) database. The NESARC database has been developed using face-to-face interviews on 43,093 adults representative of the civilian non-institutionalized population residing in the United States. The lifetime prevalence of any MI (mood, anxiety, psychotic, or personality disorder) in the NESARC sample was 33.7 %, while 66.2 % had no lifetime MI. Rates of transition from substance use to a SUD were calculated as the prevalence of lifetime SUD among individuals with lifetime exposure to a substance. The rate of transition from substance use to SUD was higher for individuals with a lifetime diagnosis of MI than for individuals without any MI. This was true for cocaine (OR = 1.83), cannabis (OR = 1.91), nicotine (OR = 3.24), alcohol (OR = 2.2), and hallucinogens (OR = 2.14) compared to controls (Lev-Ran et al. 2013). Interestingly, the SUD liability for nicotine was highest among the substances studied, in the range of 60 % for individuals with mood, anxiety, personality, and psychotic disorders (Lev-Ran et al. 2013).

Drug addiction also appears to be associated with high prevalence of tobacco use. Smoking rates ranging from 71 to 97 % were found among alcohol and other drug-dependent populations (Battjes 1988). Drug-abuse patterns seem to involve progression from one class of legal drug (alcohol or cigarettes) to marijuana/alcohol and eventually psychostimulants as well as opiate drugs (SAMHSA 2012; Yamaguchi and Kandel 1984). It has been noticed in the literature that two-thirds of drug abusers are regular tobacco smokers, a rate more than triple that of the rest of the population (Zickler 2000). In fact, early tobacco smoking has been proposed to constitute a "gateway" or rather a "common liability" (Vanyukov et al. 2012) toward substance abuse as has been suggested in the literature (Kandel and Faust 1975; Yamaguchi and Kandel 1984). Nicotine is an easily available drug for adolescents and might constitute in some cases the first contact with an addictive substance; however, some studies have pointed out that the reverse process or, as it has been denominated, the "reverse gateway theory," might also happen. The reverse gateway theory suggests that earlier regular use of other substances, as cannabis, might predict later tobacco initiation and/or nicotine dependence in those who did not use tobacco before (Peters et al. 2012). An example illustrating this reverse gateway theory is the observation of substantial percentages of studied populations found to consume marijuana prior the use of licit drugs (Tarter et al. 2006). Both scenarios (i.e., gateway and reverse gateway theories) might result in a net increase on tobacco consumption, independently of tobacco being the initial abused substance or being initiated following other addiction. Due to the high rates of polydrug use of the alcohol/tobacco combination, it has been hypothesized in the literature the possibility of mutual potentiating effects of each of these legal substances on each other.

Evidence from multiple research studies supports the existence of developmental stages and sequences in drug use (Kandel 1975). The likelihood of first initiating tobacco or other legal drugs before using illegal drugs is much greater than the opposite process. There are reports showing progression from "soft" drugs to "hard" drugs on a 75–80 % of cases versus a 20–25 % progression from "hard" to "soft" drugs of the cases depending on the sample studied (George and Moselhy 2005; Tarter et al. 2006). Accordingly, in a recent study, it was found that first initiating tobacco appeared 17.6 times greater than the likelihood of initiating cannabis (Mayet et al. 2011). Similarly to the net increase in tobacco intake that might happen in the process of escalating drugs or the potentiating effects during polydrug use mentioned above, the reverse gateway theory proposes that net increases in tobacco or other legal drugs consumption might happen following the exposure to other illegal drugs. Recent epidemiological evidence shows that alcohol and illicit drug dependence are associated with increased risk for cigarette smoking (Redner et al. 2014). Other epidemiological studies have also shown cannabis use prior tobacco consumption in a number of cases (Agrawal et al. 2012; Tullis et al. 2003; Vaughn et al. 2008). The use of cannabis during young adulthood has also been associated with increased risk of later initiation of tobacco use and nicotine dependence (Patton et al. 2005). However, in other studies, the use of marijuana in adolescence has been found only modestly associated with daily cigarette smoking and nicotine dependence in young adulthood (Timberlake et al. 2007).

There are several possibilities as to why these comorbidities exist. Here, we will focus on the behavioral data generated using animal models that could explain the cause of those comorbidities. We will notably cover the evidence collected using animal models exploring how addictive disorders, schizophrenia, mood and anxiety disorders could facilitate nicotine addiction. It appears that the reinforcing/rewarding effects of nicotine are mediated through the nicotinic acetylcholine receptors (nAchRs). nAchRs are ligand-gated ion channels composed of five subunits, labeled α_2 to α_{10} and β_2 to β_4 in the central nervous system. The combination of subunits determines the functionality of the receptor. In the brain, the most prevalent of these are the homo-oligomeric α_7 and the heteromeric $\alpha_4\beta_2^*$ nAChRs. The role and function of these combinations are not yet fully elucidated. There is clear evidence implicating $\alpha_4\beta_2^*$ nAChRs in nicotine addiction (Picciotto et al. 1998; Tapper et al. 2004). Midbrain $\alpha_4\beta_2^*$ nAChRs are critical for nicotine's ability to increase dopamine levels in the nucleus accumbens (Corrigall 1991; Dani and De Biasi 2001; Laviolette and van der Kooy 2004; Picciotto and Corrigall 2002; Watkins et al. 2000), a feature that appears critical to mediate reinforcing/rewarding effects of nicotine (see chapter entitled The Role of Mesoaccumbens Dopamine in Nicotine Dependence; this volume). It appears that other nAChRs subunits such as α_5 (Fowler et al. 2011, 2013), α_6 (Pons et al. 2008), and α_7 (Besson et al. 2012) may be implicated as well.

2 Addictive Disorder as Vulnerability for Nicotine Addiction

Animal models of drug abuse may serve to test in a relatively simple manner both gateway and reverse gateway hypotheses. Indeed, those hypotheses predict that experience with one drug can enhance the addictive properties of a different drug. Thus, by exposing experimental animals to one drug and later evaluating changes in the addictive properties of a different drug (e.g., self-administration pattern) seems a straightforward manner to test these hypotheses.

The number of studies evaluating the effects of nicotine pre-exposure on the addictive properties of other drugs is clearly superior to the number of studies evaluating the opposite process. Only a few studies have explored the effect of pre-exposure to different drugs on nicotine's-addictive properties. Thus, in a recent study, pre-exposure to tetrahydrocannabinol (THC) was able to increase the ratio of animals self-administering nicotine from 65 %, in vehicle-exposed rats, to 94 % in the THC-exposed animals (Panlilio et al. 2013). Moreover, rats pre-exposed to THC exhibited higher rates of nicotine taking over the course of acquisition training when compared to the control group. The substantial increase on the ratio of animals self-administering nicotine observed was likely by an increase in nicotine-reinforcing properties, as measured by a behavioral-economics procedure, where pre-exposure to THC produced a more persistent nicotine self-administration as the price (number of responses required) for nicotine raised (Panlilio et al. 2013). Previous studies have shown that acute pre-treatment with cannabinoid agonists and antagonists is able to modify other responses to nicotine as nicotine-induced changes in locomotor activity without changes in nicotine-evoked dopamine release (Kelsey and Calabro 2008; Rodvelt et al. 2007). Therefore, it would be interesting to further investigate the neurobiological correlates for nicotine response following pre-exposure to THC.

A recent study in mice found that pre-exposure to cocaine did not alter subsequent locomotor response to nicotine (Levine et al. 2011) (while nicotine pre-exposure facilitated subsequent responses to cocaine). In contrast, in male Sprague Dawley rats, animals receiving daily injections of cocaine for 14 days were more reactive to nicotine as compared to rats pretreated with a daily injection of saline. It should be noted that this effect was noted after 3 and 7 days of withdrawal from cocaine treatment (but not at the first day of withdrawal) (Szabo et al. 2014).

Preclinical studies have shown that a combination of legal drugs during early stages of development might also contribute to a later vulnerability to nicotine dependence. Thus, a combination of alcohol and nicotine during gestational periods might increase nicotine's-reinforcing properties during adulthood (Matta and Elberger 2007). Interestingly, alcohol-naïve offspring of rats selectively bred for high alcohol intake exhibited increased vulnerability to nicotine self-administration and relapse (Le et al. 2006). Conversely, it was previously reported that rats voluntarily consume nicotine or alcohol independently of each other, and the pre-exposure to either nicotine or alcohol did not affect subsequent intake of both drugs in combination (Marshall et al. 2003). This issue has been explored more recently using procedure in which rats voluntarily self-administer nicotine, alcohol, or both at different time points of the experimental procedure (Le et al. 2010). It appears that the nicotine intake in overall was similar in animals that got previously access to alcohol and that were trained secondarily to self-administer nicotine, as compared to animals that were trained directly to self-administer nicotine without previous alcohol access (Le et al. 2010).

3 Schizophrenia as Vulnerability for Nicotine Addiction

There are over 60 animal models of schizophrenia, and they all assess varying aspects of the disorder. Some of them model the positive symptoms, which include hallucinations and delusions. Others model the negative symptoms such as avolition and anhedonia, including changes in social functioning. Finally, some animal models study the cognitive symptoms such as impaired memory and attention. Very limited work has been conducted exploring the responses to nicotine on those models.

3.1 The Dopamine Transporter Knockout Mouse

Dopamine transporter (DAT) knockout mice (Giros et al. 1996) present a hyperdopaminergic phenotype that has been deemed relevant for schizophrenia. These mice exhibit behaviors such as perseverative locomotor activity, stereotypy, and cognitive and behavioral inflexibility. DAT knockout mice were shown to have deficits in cognitive tasks (the cued and spatial versions of the Morris water maze) and administration of nicotine improved performance such that latencies, distance travelled, and successful trials needed to reach the platform approached levels seen in the wild type (Weiss et al. 2007).

3.2 Neurodevelopmental Models of Schizophrenia

The neonatal quinpirole and neonatal ventral hippocampal lesion models provide a means to assess some of the symptoms that are believed to be caused by neurodevelopmental factors. In the neonatal quinpirole model (Brown et al. 2012), quinpirole is administered to rats neonatally, which produces behaviors that are consistent with symptoms of schizophrenia in humans (Cope et al. 2010). In the neonatal ventral hippocampal model, the part of the hippocampus that projects to the frontal cortex is lesioned, and deficits emerge later in adolescence and adulthood (Lipska et al. 2002; Lipska and Weinberger 2002; Sams-Dodd et al. 1997).

It has been found that in both quinpirole-treated rats and rats with neonatal ventral hippocampal lesions, locomotor sensitization to nicotine was increased (Berg and Chambers 2008; Perna et al. 2008), suggesting that these animals are more vulnerable to become addicted to nicotine. Indeed, in a subsequent study, it was found that neonatal ventral hippocampus rats took more nicotine infusions and required fewer trials to criterion than did those with sham lesions (Berg et al. 2014).

3.3 The Pre-pulse Inhibition Model of Schizophrenia

Pre-pulse inhibition (PPI) is specifically a model of cognitive impairment in schizophrenia and is perhaps the most widely used model of schizophrenia in animals. In rats, a number of studies of the effects of nicotine on PPI have been conducted. In one study, Curzon et al. (1994) found that an acute nicotine injection improved PPI (Curzon et al. 1994), a finding that is dependent on the strain of rat used (Acri et al. 1995) and the species (Schreiber et al. 2002). Indeed, chronic nicotine impaired PPI in Long Evans rats (Faraday et al. 1998) but enhanced it in Sprague Dawley rats (Faraday et al. 1999) and decreased PPI in Sprague Dawley rats (Schreiber et al. 2002). The reasons for these discrepancies are unknown, but together, they point to the fact that nicotine may improve cognition. In another study, rats were given daily injections with nicotine and their response to nicotine analyzed in terms of basal levels of PPI (Kayir et al. 2011). In rats that exhibited low-baseline PPI, more sensitization to nicotine was observed than in those with high basal levels of PPI.

Taken together, there are so far relatively few investigations that have explored how animal models of schizophrenia impact nicotine's effects. The existing evidence suggests that locomotor, reinforcing, and possibly cognitive effects of nicotine are enhanced in some of those animal models. However, those data should be considered preliminary as more data is needed to be able to conclude with certainty.

4 Mood Disorders as Vulnerability for Nicotine Addiction

Depression-like phenotypes are most commonly assessed in rodents using the forced swim test or the sucrose preference test. In the forced swim test, rodents are placed in a beaker of water from which they cannot escape. At first, rodents will struggle and swim around. After a period, they will become immobile and float. Greater time spent immobile during the test is thought to reflect "despair" and the forced swim test is often used to screen for antidepressant medications as drugs with known antidepressant properties tend to decrease time spent immobile (Porsolt et al. 1978, 1977). The sucrose preference test is meant to measure anhedonia or decreased ability to find pleasure in normally pleasurable things. Rodents are given free choice between drinking from a bottle of water or a bottle of sucrose water available in the same cage. Usually, rodents show a strong preference for sucrose water; a significant decrease or lack of such a preference is interpreted as anhedonia (Katz 1982; Nestler and Hyman 2010).

4.1 Nicotine Effects in Untreated Rodents

Some studies have found decreased immobility time in the forced swim test in naïve, untreated rodents. Whether administered acutely, subchronically, or chronically, nicotine was found to decrease immobility in adult male Wistar rats, without evidence of post-chronic withdrawal (Vázquez-Palacios et al. 2004). Interestingly, subchronic nicotine exposure also appeared to synergize with the antidepressant effects of fluoxetine in this study. However, other studies have also found no effects of nicotine on forced swim test performance, suggesting sex and strain differences greatly affect response to nicotine (Tizabi et al. 1999, 2009, 2010). There may be age-dependent effects for nicotine's antidepressant effects also, as acute nicotine had no effect in adolescent Sprague-Dawley rats despite finding them in adults (Villégier et al. 2010). Acute nicotine has also been found to decrease immobility in mice (Andreasen et al. 2009; Andreasen and Redrobe 2009a, b; Suemaru et al. 2006).

4.2 Nicotine Effects in Rat Strains that Model Depression

Nicotine has antidepressant effects in three inbred rat lines that show depression-like phenotypes. The flinders sensitive line (FSL) of rats was selectively bred to be hypersensitive to cholinergic stimulation and spend more time immobile in the forced swim test than the flinders resistant line (FRL) (Tizabi et al. 1999). Tizabi and colleagues found that acute or chronic (14 day) nicotine treatment significantly decreased immobility time. Wistar-Kyoto (WKY) rats are another inbred line. In addition to greater immobility in the forced swim test, they have altered sleep patterns, and more anxious and prone to stress ulcers (Tizabi et al. 2010). Adult female WKY rats along with control Wistar rats were given nicotine once or twice a day for 14 days and then were tested on the forced swim test at two time periods after the last injection (Tizabi et al. 2010). When tested 15 min after the last injection, once- or twice-daily nicotine increased immobility in WKY rats but not controls. However, when tested 18 h after the last injection, twice-daily treatment led to decreased immobility in WKY rats. Finally, Fawn-Hooded (FH) rats show depressive-like behaviors, voluntarily consume high alcohol contents and also show greater immobility in the forced swim test. Similar to the FSL rats, both acute and chronic nicotine treatments decreased immobility time in FH rats (Tizabi et al. 2009).

4.3 Nicotine Effects in Environmentally Induced Models of Depression

Antidepressant effects for nicotine have also been demonstrated in several animal models of depression that depend on environmental manipulations. Chronic nicotine had antidepressant effects in a learned helplessness model (Semba et al. 1998). In this manipulation, rats were placed in a cage where foot shocks are first delivered in an escapable fashion. Rats could stop the shock by pulling and releasing a disk mounted on the ceiling. This was followed by a second phase of inescapable shocks. Two days after the inescapable shock, rats were tested again in an escapable manner. Rats that failed to escape during these post-shock tests were considered as learned helpless. 14 days of nicotine treatment significantly decreased escape failures. Importantly, this antidepressant-like effect could be blocked by co-treatment with mecamylamine, a nicotinic receptor antagonist.

Rat neonates given clomipramine, a monoamine re-uptake inhibitor, show depressive-like features such as decreased pleasure seeking, altered sleep, and greater immobility in the forced swim test. Acute, subchronic, and chronic administrations of nicotine reduced immobility in neonatal clomipramine-treated rats (Vázquez-Palacios et al. 2005). Chronic mild stress is yet another environmental manipulation used to model depression-like phenotypes; it decreases sucrose preference and impairs working memory (Andreasen et al. 2011). Nicotine reversed such decreased sucrose preference in a manner equal to the antidepressant sertraline (Andreasen et al. 2011).

In recent literature, a study in the olfactory bulbectomy (OBX) model of depression arguably provides the most relevant test to date of the self-medication hypothesis (Vieyra-Reyes et al. 2008). The authors allowed OBX rats and sham-operated controls to voluntarily intake oral nicotine for 14 days or injected them daily with nicotine for 14 days before measuring them on the forced swim test. The effects of nicotine were compared to transcranial magnetic stimulation. Nicotine, whether taken orally in a self-regulated manner or via daily injections, decreased depression-like symptoms more than transcranial magnetic stimulation. Self-regulated oral nicotine decreased immobility in the forced swim test in both sham-operated and OBX rats. Interestingly, OBX animals had greater self-regulated intake of nicotine than sham-operated animals, which can be consistent with the self-medication hypothesis. A more recent intravenous nicotine self-administration study in a heterogenous stock of outbred rats also found that measures of depression predicted nicotine intake (Wang et al. 2014).

5 Anxiety Disorders as Vulnerability for Nicotine Addiction

Two behavioral tests have been used most frequently to model anxiety disorder-like states in rodents: the social interaction test and the elevated plus maze (Pellow et al. 1985). The social interaction test is thought to model generalized anxiety disorder and measures social interaction behaviors such as following, sniffing, and grooming as well as aggression toward a conspecific. Anxious rodents will tend to spend significantly less time in social interaction with a partner. The elevated plus maze is an elevated platform with four arms. Two opposing arms are "closed" with high arms, and the other two arms are open with no walls. The primary measures are percent time spent in open arms and number of open-arm entries. Most rodents will prefer spending exploring in the closed arms than the more threatening open arms. Decreased entries into and time spent in the open arms are considered a measure of anxiety. There can be two trials associated with the elevated plus maze. A first 5 min exposure is thought to model the escape components of panic disorder, whereas a second 5 min exposure is thought to model specific phobia (File et al. 2000).

5.1 Acute Systemic Nicotine and Anxiety

The effect of acute systemic nicotine on behavior in the social interaction test is highly dependent on dose and timing of testing. Low doses tended to be anxiolytic, whereas high doses are anxiogenic (File et al. 1998). Low doses could also have anxiolytic or anxiogenic responses depending on the timing. Testing 5 min after a low dose produced anxiogenic effects, whereas testing 30 min after injection was anxiolytic and testing 60 min after injection was again anxiogenic (Irvine et al. 1999).

Acute systemic nicotine has even more complex effects in the elevated plus maze that are not dose-dependent but probably dependent on a mix of factors including age, sex, strain, and basal anxiety levels. Anxiolytic (Villégier et al. 2010), anxiogenic (Zarrindast et al. 2012), and no effects of nicotine (Braun et al. 2011) have been found in the elevated plus maze in rats. Nicotine had anxiolytic effects in adolescent male rats but was anxiogenic for adult males and females whether adolescent or adult (Elliott et al. 2004). Bidirectional responses have also been found in mice, with some investigators reporting anxiolytic effects with a low-dose and anxiogenic effects with high doses in CD1 mice (Balerio et al. 2005, 2006) and Swiss Webster mice (Varani et al. 2012), while others have found primarily anxiogenic effects at an intermediate dose in Swiss mice (Biala and Kruk 2009; Biala et al. 2009).

5.2 Chronic Nicotine and Anxiety

Perhaps of more relevance, to the question of whether people with anxiety disorder smoke to self-medicate, are studies on longer term dosing regimens of nicotine. Irvine et al. (2001) found tolerance to nicotine's effects on anxiety with longer courses of administration in the social interaction test. They found nicotine had anxiogenic effects 30 min after an acute dose. However, if this same dose was administered for a week, this anxiogenic effect was abolished by the seventh day (Irvine et al. 2001). Interestingly, after 7 days of low-dose nicotine, a novel anxiolytic effect was found when subjects were tested 5 min post-injection. However, tolerance also developed to this anxiolytic effect with a further 7 days of low-dose nicotine. In the elevated plus maze, chronic nicotine also leads to tolerance to anxiogenic effects but sensitization to anxiolytic effects (Bhattacharya et al. 1995; Ericson et al. 2000; Irvine et al. 2001; Olausson et al. 2001). Tolerance to the acute anxiogenic effects of nicotine were also found in mice after a 7-day subchronic treatment regime (Biala and Kruk 2009; Biala et al. 2009) but in the elevated plus maze.

Of most relevance to the question of self-medication, Irvine et al. (2000) measured anxiety using the social interaction test in rats that were self-administering nicotine. They found that rats on self-administering nicotine had significantly less social interactions with no reductions in locomotion, suggesting an anxiogenic effect of nicotine. These results are contrary to the hypothesis that smokers smoke for the anxiolytic effects of nicotine. However, a recent self-administration study in outbred rats also found that measures of anxiety could help predict nicotine intake and reinstatement after extinction in male rats (Wang et al. 2014).

Studies show that withdrawal from chronic nicotine can increase anxiety. In the social interaction test, rats withdrawn for 72 h after one or two weeks of nicotine injection showed anxiogenic responses (Irvine et al. 1999). Interestingly, this effect seems specific to nicotine injections, as rats tested 24 or 72 h after 4 weeks of nicotine self-administration did not show an anxiogenic effect (Irvine et al. 2000). In the elevated plus maze as well, withdrawal after chronic nicotine injections is anxiogenic (Bhattacharya et al. 1995; Irvine et al. 2001).

5.3 Nicotine in Development and Anxiety

Recent studies looking at early-life exposure to nicotine also demonstrate significant anxiogenic effects but also in complex sex- and strain-dependent manners. Prenatal nicotine exposure in CD1 mice led to greater anxiety in the elevated platform and Suok tests but not in the elevated plus maze (Santiago and Huffman 2014). Prenatal and postnatal nicotine exposure to nicotine reduced time spent in open arms in C57BL/6 male mice as adults but increased open-arm time in DBA/2J female adult mice (Balsevich et al. 2014). In rats, adolescent nicotine exposure led to decreased exploration in the open field and quicker retreat to the perimeter of the field, suggesting heightened anxiety (Slawecki et al. 2003, 2004). However, another study found that adolescent nicotine exposure did not have any effects on elevated plus maze behavior as adults (Kupferschmidt et al. 2010).

6 Conclusions

Despite the high prevalence of tobacco smoking in all those psychiatric conditions, very few studies have explored using animal models how the reinforcing effects are altered, concerning the issue of drug addiction enhancing the vulnerability for nicotine addiction. Very few studies have been performed, and recent evidence suggests that THC exposure might increase the reinforcing effects of nicotine, but the situation is less clear following alcohol or cocaine exposure. Further studies are needed in this area, concerning the link with schizophrenia and nicotine addiction. Very few studies have been conducted and not using the most relevant models for most of those studies. It appears that animal models of schizophrenia may enhance nicotine-addictive effects, but the evidence collected so far is not clear enough to be able to conclude. In addition to the known antidepressant effects of nicotine, there is also evidence that animals more prone to "depressive states" may have higher nicotine intake. The situation appears more complicated for anxiety as nicotine has sometimes opposite effects on anxiety (bimodal modulation), but there seems to be an enhanced nicotine intake under those models. Taken together, this review indicates that there is a need for more studies to validate the findings in this area and that some of those studies may support the fact that reinforcing effects of nicotine may be enhanced in those animal models of psychiatric disorders.

Acknowledgments *Funding* Dr. Le Foll's research is supported by CAMH, the Campbell Family Mental Health Research Institute, CIHR and the National Institute on Drug Abuse at the National Institutes of Health (1R21DA033515-01).

References

Acri JB, Brown KJ, Saah MI, Grunberg NE (1995) Strain and age differences in acoustic startle responses and effects of nicotine in rats. Pharmacol Biochem Behav 50:191–198

Agrawal A, Budney AJ, Lynskey MT (2012) The co-occurring use and misuse of cannabis and tobacco: a review. Addiction 107:1221–1233

Andreasen JT, Redrobe JP (2009a) Antidepressant-like effects of nicotine and mecamylamine in the mouse forced swim and tail suspension tests: role of strain, test and sex. Behav Pharmacol 20:286–295

Andreasen JT, Redrobe JP (2009b) Nicotine, but not mecamylamine, enhances antidepressant-like effects of citalopram and reboxetine in the mouse forced swim and tail suspension tests. Behav Brain Res 197:150–156

Andreasen JT, Olsen GM, Wiborg O, Redrobe JP (2009) Antidepressant-like effects of nicotinic acetylcholine receptor antagonists, but not agonists, in the mouse forced swim and mouse tail suspension tests. J Psychopharmacol 23:797–804

Andreasen JT, Henningsen K, Bate S, Christiansen S, Wiborg O (2011) Nicotine reverses anhedonic-like response and cognitive impairment in the rat chronic mild stress model of depression: comparison with sertraline. J Psychopharmacol 25:1134–1141

Balerio GN, Aso E, Maldonado R (2005) Involvement of the opioid system in the effects induced by nicotine on anxiety-like behaviour in mice. Psychopharmacology 181:260–269

Balerio GN, Aso E, Maldonado R (2006) Role of the cannabinoid system in the effects induced by nicotine on anxiety-like behaviour in mice. Psychopharmacology 184:504–513

Balsevich G, Poon A, Goldowitz D, Wilking Ja (2014) The effects of pre- and post-natal nicotine exposure and genetic background on the striatum and behavioral phenotypes in the mouse. Behav Brain Res 266:7–18

Battjes RJ (1988) Smoking as an issue in alcohol and drug abuse treatment. Addict Behav 13:225–230

Berg SA, Chambers RA (2008) Accentuated behavioral sensitization to nicotine in the neonatal ventral hippocampal lesion model of schizophrenia. Neuropharmacology 54:1201–1207

Berg SA, Sentir AM, Cooley BS, Engleman EA, Chambers RA (2014) Nicotine is more addictive, not more cognitively therapeutic in a neurodevelopmental model of schizophrenia produced by neonatal ventral hippocampal lesions. Addict Biol 19:1020–1031

Besson M, David V, Baudonnat M, Cazala P, Guilloux JP, Reperant C, Cloez-Tayarani I, Changeux JP, Gardier AM, Granon S (2012) Alpha7-nicotinic receptors modulate nicotine-induced reinforcement and extracellular dopamine outflow in the mesolimbic system in mice. Psychopharmacology 220:1–14

Bhattacharya SK, Chakrabarti A, Sandler M, Glover V (1995) Rat brain monoamine oxidase A and B inhibitory (tribulin) activity during drug withdrawal anxiety. Neurosci Lett 199:103–106

Biala G, Kruk M (2009) Effects of co-administration of bupropion and nicotine or D-amphetamine on the elevated plus maze test in mice. J Pharm Pharmacol 61:493–502

Biala G, Kruk M, Budzynska B (2009) Effects of the cannabinoid receptor ligands on anxiety-related effects of d-amphetamine and nicotine in the mouse elevated plus maze test. J Physiol Pharmacol 60:113–122

Braun Aa, Skelton MR, Vorhees CV, Williams MT (2011) Comparison of the elevated plus and elevated zero mazes in treated and untreated male Sprague-Dawley rats: effects of anxiolytic and anxiogenic agents. Pharmacol Biochem Behav 97:406–415

Brooner RK, King VL, Kidorf M, Schmidt CW Jr, Bigelow GE (1997) Psychiatric and substance use comorbidity among treatment-seeking opioid abusers. Arch Gen Psychiatry 54:71–80

Brown RW, Maple AM, Perna MK, Sheppard AB, Cope ZA, Kostrzewa RM (2012) Schizophrenia and substance abuse comorbidity: nicotine addiction and the neonatal quinpirole model. Dev Neurosci 34:140–151

Cope ZA, Huggins KN, Sheppard AB, Noel DM, Roane DS, Brown RW (2010) Neonatal quinpirole treatment enhances locomotor activation and dopamine release in the nucleus accumbens core in response to amphetamine treatment in adulthood. Synapse 64:289–300

Corrigall WA (1991) Understanding brain mechanisms in nicotine reinforcement. Br J Addict 86:507–510

Curzon P, Kim DJ, Decker MW (1994) Effect of nicotine, lobeline, and mecamylamine on sensory gating in the rat. Pharmacol Biochem Behav 49:877–882

Dani JA, De Biasi M (2001) Cellular mechanisms of nicotine addiction. Pharmacol Biochem Behav 70:439–446

Elliott BM, Faraday MM, Phillips JM, Grunberg NE (2004) Effects of nicotine on elevated plus maze and locomotor activity in male and female adolescent and adult rats. Pharmacol Biochem Behav 77:21–28

Ericson M, Olausson P, Engel JA, Söderpalm B (2000) Nicotine induces disinhibitory behavior in the rat after subchronic peripheral nicotinic acetylcholine receptor blockade. Eur J Pharmacol 397:103–111

Faraday MM, Rahman MA, Scheufele PM, Grunberg NE (1998) Nicotine administration impairs sensory gating in Long-Evans rats. Pharmacol Biochem Behav 61:281–289

Faraday MM, O'Donoghue VA, Grunberg NE (1999) Effects of nicotine and stress on startle amplitude and sensory gating depend on rat strain and sex. Pharmacol Biochem Behav 62:273–284

File SE, Kenny PJ, Ouagazzal AM (1998) Bimodal modulation by nicotine of anxiety in the social interaction test: role of the dorsal hippocampus. Behav Neurosci 112:1423–1429

File SE, Cheeta S, Kenny PJ (2000) Neurobiological mechanisms by which nicotine mediates different types of anxiety. Eur J Pharmacol 393:231–236

Fowler CD, Lu Q, Johnson PM, Marks MJ, Kenny PJ (2011) Habenular alpha5 nicotinic receptor subunit signalling controls nicotine intake. Nature 471:597–601

Fowler CD, Tuesta L, Kenny PJ (2013) Role of alpha5* nicotinic acetylcholine receptors in the effects of acute and chronic nicotine treatment on brain reward function in mice. Psychopharmacology 229:503–513

George SA, Moselhy H (2005) "Gateway hypothesis"—a preliminary evaluation of variables predicting non-conformity. Addict Disord Treat 4:39–40

Giros B, Jaber M, Jones SR, Wightman RM, Caron MG (1996) Hyperlocomotion and indifference to cocaine and amphetamine in mice lacking the dopamine transporter. Nature 379:606–612

Hien D, Zimberg S, Weisman S, First M, Ackerman S (1997) Dual diagnosis subtypes in urban substance abuse and mental health clinics. Psychiatr Serv 48:1058–1063

Irvine EE, Cheeta S, File SE (1999) Time-course of changes in the social interaction test of anxiety following acute and chronic administration of nicotine. Behav Pharmacol 10:691–697

Irvine EE, Bagnalasta M, Marcon C, Motta C, Tessari M, File SE, Chiamulera C (2000) Nicotine self-administration and withdrawal: modulation of anxiety in the social interaction test in rats. Psychopharmacology 153:315–320

Irvine EE, Cheeta S, File SE (2001) Tolerance to nicotine's effects in the elevated plus-maze and increased anxiety during withdrawal. Pharmacol Biochem Behav 68:319–325

Kandel D (1975) Stages in adolescent involvement in drug use. Science 190:912–914

Kandel D, Faust R (1975) Sequence and stages in patterns of adolescent drug use. Arch Gen Psychiatry 32:923–932

Katz RJ (1982) Animal model of depression: pharmacological sensitivity of a hedonic deficit. Pharmacol Biochem Behav 16:965–968

Kayir H, Goktalay G, Yavuz O, Uzbay TI (2011) Impact of baseline prepulse inhibition on nicotine-induced locomotor sensitization in rats. Behav Brain Res 216:275–280

Kelsey JE, Calabro S (2008) Rimonabant blocks the expression but not the development of locomotor sensitization to nicotine in rats. Psychopharmacology 198:461–466

Kessler DA, Hass AE, Feiden KL, Lumpkin M, Temple R (1996a) Approval of new drugs in the United States. Comparison with the United Kingdom, Germany, and Japan. JAMA 276:1826–1831

Kessler DA, Witt AM, Barnett PS, Zeller MR, Natanblut SL, Wilkenfeld JP, Lorraine CC, Thompson LJ, Schultz WB (1996b) The food and drug administration's regulation of tobacco products. N Engl J Med 335:988–994

Kupferschmidt DA, Funk D, Erb S, Lê AD (2010) Age-related effects of acute nicotine on behavioural and neuronal measures of anxiety. Behav Brain Res 213:288–292

Laviolette SR, van der Kooy D (2004) The neurobiology of nicotine addiction: bridging the gap from molecules to behaviour. Nat Rev Neurosci 5:55–65

Le AD, Li Z, Funk D, Shram M, Li TK, Shaham Y (2006) Increased vulnerability to nicotine self-administration and relapse in alcohol-naive offspring of rats selectively bred for high alcohol intake. J Neurosci 26:1872–1879

Le AD, Lo S, Harding S, Juzytsch W, Marinelli PW, Funk D (2010) Coadministration of intravenous nicotine and oral alcohol in rats. Psychopharmacology 208:475–486

Levine A, Huang Y, Drisaldi B, Griffin EA, Jr., Pollak DD, Xu S, Yin D, Schaffran C, Kandel DB, Kandel ER (2011) Molecular mechanism for a gateway drug: epigenetic changes initiated by nicotine prime gene expression by cocaine. Sci Transl Med 3:107ra109

Lev-Ran S, Imtiaz S, Rehm J, Le Foll B (2013) Exploring the association between lifetime prevalence of mental illness and transition from substance use to substance use disorders: results from the national epidemiologic survey of alcohol and related conditions (NESARC). Am J Addict 22:93–98

Lipska BK, Weinberger DR (2002) A neurodevelopmental model of schizophrenia: neonatal disconnection of the hippocampus. Neurotox Res 4:469–475

Lipska BK, Aultman JM, Verma A, Weinberger DR, Moghaddam B (2002) Neonatal damage of the ventral hippocampus impairs working memory in the rat. Neuropsychopharmacology 27:47–54

Marshall CE, Dadmarz M, Hofford JM, Gottheil E, Vogel WH (2003) Self-administration of both ethanol and nicotine in rats. Pharmacology 67:143–149

Matta SG, Elberger AJ (2007) Combined exposure to nicotine and ethanol throughout full gestation results in enhanced acquisition of nicotine self-administration in young adult rat offspring. Psychopharmacology 193:199–213

Mayet A, Legleye S, Chau N, Falissard B (2011) Transitions between tobacco and cannabis use among adolescents: a multi-state modeling of progression from onset to daily use. Addict Behav 36:1101–1105

Nestler EJ, Hyman SE (2010) Animal models of neuropsychiatric disorders. Nat Neurosci 13:1161–1169

Olausson P, Akesson P, Petersson A, Engel JA, Söderpalm B (2001) Behavioral and neurochemical consequences of repeated nicotine treatment in the serotonin-depleted rat. Psychopharmacology 155:348–361

Panlilio LV, Zanettini C, Barnes C, Solinas M, Goldberg SR (2013) Prior exposure to THC increases the addictive effects of nicotine in rats. Neuropsychopharmacology 38:1198–1208

Patton GC, Coffey C, Carlin JB, Sawyer SM, Lynskey M (2005) Reverse gateways? Frequent cannabis use as a predictor of tobacco initiation and nicotine dependence. Addiction 100:1518–1525

Pellow S, Chopin P, File SE, Briley M (1985) Validation of open:closed arm entries in an elevated plus-maze as a measure of anxiety in the rat. J Neurosci Methods 14:149–167

Perna MK, Cope ZA, Maple AM, Longacre ID, Correll JA, Brown RW (2008) Nicotine sensitization in adult male and female rats quinpirole-primed as neonates. Psychopharmacology 199:67–75

Peters EN, Budney AJ, Carroll KM (2012) Clinical correlates of co-occurring cannabis and tobacco use: a systematic review. Addiction 107:1404–1417

Picciotto MR, Corrigall WA (2002) Neuronal systems underlying behaviors related to nicotine addiction: neural circuits and molecular genetics. J Neurosci 22:3338–3341

Picciotto MR, Zoli M, Rimondini R, Lena C, Marubio LM, Pich EM, Fuxe K, Changeux JP (1998) Acetylcholine receptors containing the beta2 subunit are involved in the reinforcing properties of nicotine. Nature 391:173–177

Pons S, Fattore L, Cossu G, Tolu S, Porcu E, McIntosh JM, Changeux JP, Maskos U, Fratta W (2008) Crucial role of alpha4 and alpha6 nicotinic acetylcholine receptor subunits from ventral tegmental area in systemic nicotine self-administration. J Neurosci 28:12318–12327

Porsolt RD, Bertin A, Jalfre M (1977) Behavioral despair in mice: a primary screening test for antidepressants. Arch Int Pharmacodyn Ther 229:327–336

Porsolt RD, Anton G, Blavet N, Jalfre M (1978) Behavioral despair in rats: a new model sensitive to antidepressant treatments. Eur J Pharmacol 47:379–391

Redner R, White TJ, Harder VS, Higgins ST (2014) Vulnerability to smokeless tobacco use among those dependent on alcohol or illicit drugs. Nicotine Tob Res 16:216–223

Regier DA, Farmer ME, Rae DS, Locke BZ, Keith SJ, Judd LL, Goodwin FK (1990) Comorbidity of mental disorders with alcohol and other drug abuse. Results from the epidemiologic catchment area (ECA) study. JAMA 264:2511–2518

Rodvelt KR, Bumgarner DM, Putnam WC, Miller DK (2007) WIN-55,212-2 and SR-141716A alter nicotine-induced changes in locomotor activity, but do not alter nicotine-evoked [3H] dopamine release. Life Sci 80:337–344

SAMHSA (2012) Results from the 2012 national survey on drug use and health: summary of national findings, SAMHSA, Rockville

Sams-Dodd F, Lipska BK, Weinberger DR (1997) Neonatal lesions of the rat ventral hippocampus result in hyperlocomotion and deficits in social behaviour in adulthood. Psychopharmacology 132:303–310

Santiago SE, Huffman KJ (2014) Prenatal nicotine exposure increases anxiety and modifies sensorimotor integration behaviors in adult female mice. Neurosci Res 79:41–51

Schreiber R, Dalmus M, De Vry J (2002) Effects of alpha 4/beta 2- and alpha 7-nicotine acetylcholine receptor agonists on prepulse inhibition of the acoustic startle response in rats and mice. Psychopharmacology 159:248–257

Semba J, Mataki C, Yamada S, Nankai M, Toru M (1998) Antidepressantlike effects of chronic nicotine on learned helplessness paradigm in rats. Biol Psychiatry 43:389–391

Slawecki CJ, Gilder A, Roth J, Ehlers CL (2003) Increased anxiety-like behavior in adult rats exposed to nicotine as adolescents. Pharmacol Biochem Behav 75:355–361

Slawecki CJ, Thorsell A, Ehlers CL (2004) Long-term neurobehavioral effects of alcohol or nicotine exposure in adolescent animal models. Ann NY Acad Sci 1021:448–452

Suemaru K, Yasuda K, Cui R, Li B, Umeda K (2006) Antidepressant-like action of nicotine in forced swimming test and brain serotonin in mice. Physiol Behav 88:545–549

Swendsen J, Conway KP, Degenhardt L, Glantz M, Jin R, Merikangas KR, Sampson N, Kessler RC (2010) Mental disorders as risk factors for substance use, abuse and dependence: results from the 10-year follow-up of the national comorbidity survey. Addiction 105:1117–1128

Szabo ST, Fowler JC, Froeliger B, Lee TH (2014) Time-dependent changes in nicotine behavioral responsivity during early withdrawal from chronic cocaine administration and attenuation of cocaine sensitization by mecamylamine. Behav Brain Res 262:42–46

Tapper AR, McKinney SL, Nashmi R, Schwarz J, Deshpande P, Labarca C, Whiteaker P, Marks MJ, Collins AC, Lester HA (2004) Nicotine activation of alpha4* receptors: sufficient for reward, tolerance, and sensitization. Science 306:1029–1032

Tarter RE, Vanyukov M, Kirisci L, Reynolds M, Clark DB (2006) Predictors of marijuana use in adolescents before and after licit drug use: examination of the gateway hypothesis. Am J psychiatry 163:2134–2140

Timberlake DS, Haberstick BC, Hopfer CJ, Bricker J, Sakai JT, Lessem JM, Hewitt JK (2007) Progression from marijuana use to daily smoking and nicotine dependence in a national sample of U.S. adolescents. Drug Alcohol Depend 88:272–281

Tizabi Y, Overstreet DH, Rezvani AH, Louis VA, Clark E, Janowsky DS, Kling MA (1999) Antidepressant effects of nicotine in an animal model of depression. Psychopharmacology 142:193–199

Tizabi Y, Getachew B, Rezvani AH, Hauser SR, Overstreet DH (2009) Antidepressant-like effects of nicotine and reduced nicotinic receptor binding in the Fawn-Hooded rat, an animal model of co-morbid depression and alcoholism. Prog Neuropsychopharmacol Biol Psychiatry 33:398–402

Tizabi Y, Hauser SR, Tyler KY, Getachew B, Madani R, Sharma Y, Manaye KF (2010) Effects of nicotine on depressive-like behavior and hippocampal volume of female WKY rats. Prog Neuropsychopharmacol Biol Psychiatry 34:62–69

Tullis LM, Dupont R, Frost-Pineda K, Gold MS (2003) Marijuana and tobacco: a major connection? J Addict Dis 22:51–62

Vanyukov MM, Tarter RE, Kirillova GP, Kirisci L, Reynolds MD, Kreek MJ, Conway KP, Maher BS, Iacono WG, Bierut L, Neale MC, Clark DB, Ridenour TA (2012) Common liability to addiction and "gateway hypothesis": theoretical, empirical and evolutionary perspective. Drug Alcohol Depend 123:S3–S17

Varani AP, Moutinho LM, Bettler B, Balerio GN (2012) Acute behavioural responses to nicotine and nicotine withdrawal syndrome are modified in GABA(B1) knockout mice. Neuropharmacology 63:863–872

Vaughn M, Wallace J, Perron B, Copeland V, Howard M (2008) Does marijuana use serve as a gateway to cigarette use for high-risk African-American youth? Am J Drug Alcohol Abuse 34:782–791

Vázquez-Palacios G, Bonilla-Jaime H, Velázquez-Moctezuma J (2004) Antidepressant-like effects of the acute and chronic administration of nicotine in the rat forced swimming test and its interaction with fluoxetine [correction of flouxetine]. Pharmacol Biochem Behav 78:165–169

Vázquez-Palacios G, Bonilla-Jaime H, Velázquez-Moctezuma J (2005) Antidepressant effects of nicotine and fluoxetine in an animal model of depression induced by neonatal treatment with clomipramine. Prog Neuropsychopharmacol Biol Psychiatry 29:39–46

Vieyra-Reyes P, Mineur YS, Picciotto MR, Túnez I, Vidaltamayo R, Drucker-Colín R (2008) Antidepressant-like effects of nicotine and transcranial magnetic stimulation in the olfactory bulbectomy rat model of depression. Brain Res Bull 77:13–18

Villégier A-S, Gallager B, Heston J, Belluzzi JD, Leslie FM (2010) Age influences the effects of nicotine and monoamine oxidase inhibition on mood-related behaviors in rats. Psychopharmacology 208:593–601

Wang T, Han W, Wang B, Jiang Q, Solberg-Woods LC, Palmer AA, Chen H (2014) Propensity for social interaction predicts nicotine-reinforced behaviors in outbred rats. Genes Brain Behav 13:202–212

Watkins SS, Koob GF, Markou A (2000) Neural mechanisms underlying nicotine addiction: acute positive reinforcement and withdrawal. Nicotine Tob Res 2:19–37

Weiss S, Nosten-Bertrand M, McIntosh JM, Giros B, Martres MP (2007) Nicotine improves cognitive deficits of dopamine transporter knockout mice without long-term tolerance. Neuropsychopharmacology 32:2465–2478

Yamaguchi K, Kandel DB (1984) Patterns of drug use from adolescence to young adulthood: II. Sequences of progression. Am J Public Health 74:668–672

Zarrindast MR, Aghamohammadi-Sereshki A, Rezayof A, Rostami P (2012) Nicotine-induced anxiogenic-like behaviours of rats in the elevated plus-maze: possible role of NMDA receptors of the central amygdala. J Psychopharmacol 26:555–563

Zickler P (2000) Nicotine craving and heavy smoking may contribute to increased use of cocaine and heroin. NIDA Notes 15

Index

D.J.K. Balfour and M.R. Munafò (eds.), *The Neuropharmacology of Nicotine Dependence*, Current Topics in Behavioral Neurosciences 24,
DOI 10.1007/978-3-319-13482-6

The manufacturer's authorised representative in the EU is Springer Nature Customer Service Centre GmbH, Europaplatz 3, 69115 Heidelberg, Germany. If you have any concerns regarding our products, please contact ProductSafety@springernature.com

Printed and bound by CPI Group (UK) Ltd, Croydon, CR0 4YY
15/07/2026
02167645-0011